MILITARISM AND GLOBAL ECOLOGY

IDEAS in CONFLICT

Gary E. McCuen

GEM
GARY McCUEN
publications inc.

411 Mallalieu Drive
Hudson, Wisconsin 54016
Phone (715) 386-7113

Illustrations & Photo Credits

Center for Defense Information 105, Carol*Simpson 22, 47, 126, Daily World 38, Michael Powers 28, Richard Wright 75, Tribune Media Services 62, Ron Swanson 42, 52, 80, 87, 98, Steve Sack 17, U.S. Department of Defense 11, 68, 93, U.S. Department of Energy 112, 119

© 1993 By Gary E. McCuen Publications, Inc.
411 Mallalieu Drive, Hudson, Wisconsin 54016

(715) 386-7113

International Standard Book Number
0-86596-086-0 Printed in the United States
of America

CONTENTS

Ideas in Conflct   6

CHAPTER 3 WAR AND THE ENVIRONMENT: IDEAS IN CONFLICT

REASONING SKILL DEVELOPMENT

These activities may be used as individualized study guides for students in libraries and resource centers or as discussion catalysts in small group and classroom discussions.

This series features ideas in conflict on political, social, and moral issues. It presents counterpoints, debates, opinions, commentary, and analysis for use in libraries and classrooms. Each title in the series uses one or more of the following basic elements:

Introductions *that present an issue overview giving historic background and/or a description of the controversy.*

Counterpoints *and debates carefully chosen from publications, books, and position papers on the political right and left to help librarians and teachers respond to requests that treatment of public issues be fair and balanced.*

Symposiums *and forums that go beyond debates that can polarize and oversimplify. These present commentary from across the political spectrum that reflect how complex issues attract many shades of opinion.*

A **global** *emphasis with foreign perspectives and surveys on various moral questions and political issues that will help readers to place subject matter in a less culture-bound and ethnocentric frame of reference. In an ever-shrinking and interdependent world, understanding and cooperation are essential. Many issues are global in nature and can be effectively dealt with only by common efforts and international understanding.*

Reasoning skill *study guides and discussion activities provide ready-made tools for helping with critical reading and evaluation of content. The guides and activities deal with one or more of the following:*

RECOGNIZING AUTHOR'S POINT OF VIEW

INTERPRETING EDITORIAL CARTOONS

VALUES IN CONFLICT

WHAT IS EDITORIAL BIAS?

WHAT IS SEX BIAS?

WHAT IS POLITICAL BIAS?

WHAT IS ETHNOCENTRIC BIAS?

WHAT IS RACE BIAS?

WHAT IS RELIGIOUS BIAS?

*From across **the political spectrum** varied sources are presented for research projects and classroom discussions. Diverse opinions in the series come from magazines, newspapers, syndicated columnists, books, political speeches, foreign nations, and position papers by corporations and nonprofit institutions.*

About the Editor

Gary E. McCuen is an editor and publisher of anthologies for public libraries and curriculum materials for schools. Over the past years his publications have specialized in social, moral and political conflict. They include books, pamphlets, cassettes, tabloids, filmstrips and simulation games, many of them designed from his curriculums during 11 years of teaching junior and senior high school social studies. At present he is the editor and publisher of the *Ideas in Conflict* series and the *Editorial Forum* series.

CHAPTER 1

THE WAR ON NATURE:
AN OVERVIEW

1   THE WAR ON NATURE: AN OVERVIEW

THE MILITARY'S ASSAULT ON NATURE

Michael G. Renner

Michael G. Renner is a contributing editor to The New Economy *and senior researcher at Worldwatch Institute.*

Points to Consider:

1. How does militarism threaten global ecology even during "peacetime"?

2. What are the costs of repairing the damage?

3. How do the laws in Germany promote a military assault on the environment?

4. How can the public change the military's attitude?

Michael G. Renner, "The Military's War on the Environment", **The New Economy,** December, 1990.

"The problem in defense is how far you can go without destroying from within what you are trying to defend from without."

More than three decades ago, President Dwight D. Eisenhower warned that "the problem in defense is how far you can go without destroying from within what you are trying to defend from without." Meant as a warning against creating an all-powerful military-industrial complex, Eisenhower's statement is equally applicable to a problem the world is just beginning to grapple with: the military's war on the environment.

Modern warfare entails large-scale environmental devastation, as conflicts in Vietnam, Afghanistan, Central America, and the Persian Gulf amply demonstrate. In some cases, environmental modification has consciously been employed as a weapon. And it is generally agreed that nuclear war is the ultimate threat to the global environment.

But even in "peacetime" — preparing for war — the military contributes to resource depletion and environmental degradation, in some instances heavily. The production, testing, and maintenance of conventional, chemical, biological, and nuclear arms generates enormous quantities of toxic and radioactive substances, and contaminates the earth's soil, air and water. Keeping troops in a state of readiness imposes a heavy toll on large tracts of often-fragile land.

The military's troubling relationship with the environment is moving up on the world's agenda for a number of reasons. Military wastes are literally surfacing: the effects of decades of pervasive disregard for the environment in the nuclear weapons production complex, for example, simply cannot be hidden from public view. Suddenly vanished, the Cold War no longer provides a convenient smokescreen to deflect public concerns. As some of the superpowers' troops and bases are withdrawn from foreign lands, unpleasant discoveries are being made. Rising environmental awareness has sensitized ordinary citizens to issues neglected earlier, while a stream of shocking revelations of past environmental destruction by the military is forcing governments to acknowledge and begin to address the problems.

Ecological destruction during the Gulf War. U.S. Army photo

Making Peace with the Environment

As the East-West confrontation has faded, the environmental legacy of the Cold War is slowly being put on the agenda. The cost of repairing the damage wrought by permanent war preparation will be staggering.

For the United States, a 1988 General Accounting Office report calculated nuclear decontamination costs of $100-130 billion, or $2 million for every nuclear warhead the nation has produced. More recent assessments speak of $200 billion. Estimates for required outlays to deal with toxic wastes at U.S. military bases have skyrocketed from $500 million in 1983 to $20-40 billion today, and are certain to escalate further as additional toxic sites are identified and cleanup work progresses. In addition, U.S. forces stationed in Western Europe would need to spend at least $580 million to reduce air and water pollution at their bases.

Cleaning up Soviet bases in Czechoslovakia has been estimated to cost $2 million per site, or close to $300 million for all 132 installations. In Hungary, the cleanup bill could cost several tens of millions of dollars. These amounts are hardly

11

affordable in these financially strapped nations.

How Much?

The U.S. Department of Energy's budget for coping with contamination of the nuclear weapons complex has more than quadrupled between 1986 and 1991, to $4.3 billion, and the Department still refuses to spend huge sums on refurbishing production facilities.

Beyond the financial implications, however, are less quantifiable though by no means insignificant social costs. Those who have been exposed to crippling or fatal pollution in the name of national security are already paying a heavy price for the geopolitical rivalry of the past half-century. And the time required for decontaminating sites polluted by toxic and radioactive wastes may have to be measured in decades and generations. The most severely poisoned areas may prove impossible to "clean up" or otherwise rehabilitate. Fenced-off and unsuitable for any use, they may become "national sacrifice zones", ghastly monuments of the Cold War.

Under the Law?

The military sector has long considered itself beyond the purview of existing environmental laws and regulations. Public awareness of environmental problems generated by military activities is important if government agencies and their private contractors who inflicted the damage are to be held to greater accountability.

An assessment of environmentally relevant laws in West Germany, including those regulating land use, waste disposal, and emission of pollutants, found that virtually every one of them contained some loopholes that accorded the armed forces special privileges. And in a classic example of the fox guarding the henhouse, the Bundeswehr had the sole right from 1986 to inspect its own compliance with federal air pollution emission laws. Similar conditions are likely to exist in other countries as well.

Citizen Action

The issue of military-related pollution and degradation can unite environmental and peace movements. In the United States, the National Toxics Campaign and the Citizen's Clearinghouse for Hazardous Wastes are assisting communities in confronting military pollution in their own backyards. On the national level, they, along with other groups, are pushing for legislation making military facilities subject to the same environmental requirements and penalties as private polluters, and calling for an "Environmental Security Fund" that automatically provides monies for the investigation and cleanup of toxic contamination at military facilities. The Military Production Network, meanwhile, brings together groups across the country concerned about the effects of nuclear weapon production.

In West Germany, citizens outraged at the continuing stress from low-level flights demanded an end to the practice. In both parts of the newly unified Germany, toxic contamination, particularly at bases occupied by their former superpower patrons, is a "hot" issue. In Kazakhstan, in the (former) Soviet Union, a large and vocal grassroots movement opposed to all nuclear tests arose virtually overnight. Supported by the coal miners union, the "Nevada-Semipalatinsk" movement (as it calls itself, to underline the shared fate of Americans and Soviets) succeeded in forcing an end to all testing.

These examples show that strong public pressure can succeed in changing the military's attitude toward the environment. But to be more effective in their struggle, grassroots movements need adequate tools. One is "right-to-know" legislation as it is now being applied to U.S. companies' pollution in the civilian sphere. Piercing through the national security smoke screen, such legislation could require not only defense agencies but also military contractors to prepare detailed reports on hazardous substances handled and released by them and to make the date available to the public. Environmental impact statements for any future military projects, already required for the U.S. and Canadian armed forces, are an important way to identify potentially adverse impacts before they become a reality.

Unlike in the civilian sphere, however, it makes little sense to demand stricter emissions norms for bomb plants or more fuel-efficient tanks. The essence of all military operations is achieving a margin of superiority over real or perceived adversaries, at whatever environmental or other costs. The fundamental incompatibility of the military and the environment was brought home recently by a U.S. military base commander in a community hearing in Virginia: "We are in the business of protecting the nation, not the environment."

A world that wants to make peace with the environment cannot continue to fight wars or to sacrifice human health and the earth's ecosystems by preparing for war. Environmental quality joins a long list of solid reasons for moving toward disarmament. The destruction of our global natural support systems is a steep price to pay for militarized national "security".

2   THE WAR ON NATURE:
AN OVERVIEW

POLLUTING THE HIGH FRONTIER

Gar Smith

Gar Smith is editor of Earth Island Journal, *a quarterly publication dealing with global environmental concerns. This reading was taken from a speech given by Gar Smith at the Cape Canaveral commemoration of the Challenger shuttle disaster.*

Points to Consider:

1. What is the "Last Wilderness" concept?

2. How does a typical U.S. space shuttle launch affect global ecology?

3. Why are some recent deep space probes so potentially dangerous?

4. To what extent have we already polluted space?

Gar Smith, "Space—Earth's 'Last Wilderness'", **Earth Island Journal**, Spring 1989 Reprinted with permission of **Earth Island Journal**, 300 Broadway, Suite 28, San Francisco, CA 94133; (415) 788-3666. Earth Island Institute membership $25/yr., includes 4 issues of **Earth Island Journal.**.

The once-pristine space around our planet is now littered with more than 7000 pieces of space junk.

In the 1970s, when NASA's budget was cut back by the Nixon Administration, the military insisted on more lifting capacity to orbit its newest spy satellites. In an attempt to meet the military's needs and deliver what one NASA engineer called "more bang for the buck", the shuttle was redesigned for economy. One of the systems cut was the crew's escape system. A decision was made that military needs were more critical than human lives.

While most of us grew up with the idea of space as the "Last Frontier", Earth Island is now asking people to look at space, instead, as the "Last Wilderness". And like any wilderness, space needs to be protected — from wars, from exploitation, from pollution.

A Dirty Business

The environmental problems with space travel begin at the launching pad, when each lift-off creates clouds of choking smoke and toxic gases. A typical shuttle blast-off leaves behind about eight million pounds of water contaminated with corrosive hydrochloric acid. As the rocket bores its way into the sky, the waste cloud streaming from its flaming engine reacts chemically to devour the Earth's protective ozone shield.

As the Challenger disaster demonstrated three years ago, there's no guarantee of a perfect launch. That explosion not only killed seven Americans, it also killed a young Cuban fisherman who was hit by debris raining into the Gulf of Mexico. The casualties could have been much higher, however, had the cargo included either of the two deep space probes launched by NASA.

The Magellan probe to Venus and the Galileo probe to Jupiter are both powered by RTG (Radioactive Thermal Generator) plutonium packages. The Galileo carries 44.25 pounds of plutonium—enough to kill tens of thousands if dispersed in a Challenger-like explosion; enough to build seven atom bombs. (The Galileo probe has already caught fire once in the laboratory.)

It may sound impossible, but the fact is space is already polluted. The U.S. and other "spacefaring" nations are guilty of a kind of orbital "off-shore dumping".

Since the dawn of the Space Age, more than 15,000 objects

Cartoon by Steve Sack. Reprinted by permission of the **Star Tribune**, Minneapolis

have been chucked up into Earth's orbit: everything from spent payloads, rocket bodies and dead satellites, to clamps, lost wrenches and human wastes. More than 60 satellites have been broken apart in orbit (not counting the perfectly good weather satellite intentionally blown to pieces by the U.S. in a space weapons test). As a result, the once-pristine space around our planet is now littered with more than 7000 pieces of space junk the size of baseballs and larger, whizzing along at speeds of 9000 mph.

Some of that floating junk includes nuclear-powered satellites. Once promoted as safe, the fact is that nearly one-fifth of the 60 or so nuclear-powered missions launched by the U.S. and the (former) USSR have met with failure.

In 1964, a Navy transit satellite burned up over Madagascar and plunged into the Indian Ocean, trailing clouds of plutonium oxide through the stratosphere. Deadly plutonium-238 continued to drift down to Earth over the next six years.

In 1969, two Soviet moon missions ignited in the Earth's atmosphere, releasing detectable amounts of radioactivity. A year later, an Apollo 13 lunar lander mission missed its intended target and wound up 2,700 fathoms under the Pacific off New Zealand, with 8.6 pounds of plutonium on board. And in the most notorious case, so far, the (former) USSR's Cosmos 954 crashed into a remote region of northern Canada on January 24, 1978, spreading radioactive debris over 40,000 square miles. The cleanup took six months and cost $14 million.

The Soviets responded to the Cosmos caper with a classic techno-fix. The satellites were redesigned to jettison their nuclear cores for "safe storage" in a higher orbit. As a result, we now have a "nuclear dump" located in a region of space 600 miles over our heads filled with 3000 pounds of plutonium and uranium-238. Perhaps this is an appropriate occasion to suggest that our president nominate space for listing among the EPA's "Superfund Sites".

Meanwhile, scientists (including a team right here at the University of Florida), are working hard to place a new generation of nuclear reactors in space by 1995. These won't be the traditional low-power jobs that turn out only enough electricity to power a blow-dryer. These brutes are designed to power Star Wars weapons. The smallest would produce one megawatt of power. The largest would weigh more than 3,000 pounds and kick out 500 megawatts. (Picture an orbiting Lincoln Continental turning out enough electricity to light Miami for a year.) Current scenarios envision 100 of these Star Wars reactors cruising in Earth's orbit. (Most funding for the Star War program has now been cut by Congress.)

This presents us with a very sticky problem. On one hand, you've got this hailstorm of orbital litter zipping along at hypersonic speeds. On the other hand, you've got these guys who want to orbit nuclear reactors as big as a barn door. With space litter growing at the rate of 15 percent a year and scientists predicting there's a 1-in-10 chance of the larger satellites being hit in the next 15 years, trying to put reactors in space makes about as much sense as building a beehive on a freeway.

The "Disposable Planet"

We have to take care not to think of space travel as a convenient "escape hatch" from the problems that we've created on Earth. The "Disposable Planet Mentality" argues that we've simply outgrown this planet and now it's time to move on to the stars. With the impacts of global warming suddenly falling about our ears, this temptation is greater than ever. But before we embark on a mission to Mars, we need to undertake what former astronaut Sally Ride has called a "Mission to Earth". After all, if we still haven't learned how to survive within the limits of our planetary ecosystem, how can we possibly hope to thrive in the cramped quarters of a voyaging spaceship?

I'd like to propose today that we institute a People's Law. It would require that every would-be world leader be shot into space and left to orbit the planet for at least a week before assuming office.

We know from past experience that something wonderful and profound seems to happen to human beings once they've gazed down at our home planet from the perspective of a small, floating capsule. They get to feeling downright. . .protective. Send up a politician; down comes a poet. It couldn't hurt.

If we have a future at all it won't be found in the stars. It's got to be created right here on Earth.

3   THE WAR ON NATURE:
AN OVERVIEW

NUCLEAR WEAPONS AND
ECOLOGICAL SUICIDE

Dean Babst and Margo Schulter

Dean Babst and Margo Schulter originally wrote this article as a fourteen page report for the Nuclear Age Peace Foundation.

Points to Consider:

1. Describe the human and ecological damage from the Chernobyl nuclear disaster.

2. How would one Trident submarine attack compare to Chernobyl?

3. What aspect of nuclear tactics has been ignored by military planners?

4. How would global ecology be threatened by accidents not involving actual nuclear attack?

Dean Babst and Margo Schulter, "Nuclear Weapons: The Suicidal Defense," **Earth Island Journal,** Spring 1989. Reprinted with permission of **Earth Island Journal,** 300 Broadway, Suite 28, San Francisco, CA 94133; (415) 788-3666. Earth Island Institute membership $25/yr., includes 4 issues of **Earth Island Journal.**

Just one U.S. Trident submarine could release radioactive fallout equivalent to 377 Chernobyl power plant accidents.

It is widely known that nuclear weapons contain radioactive materials. What is not generally known is that the use of just a few modern nuclear weapons can produce a vast amount of global radioactivity. Just one U.S. Trident submarine could release radioactive fallout equivalent to 377 Chernobyl power plant accidents. Can anyone adequately imagine the destructive effect of the global fallout from more than 300 Chernobyls?

The world's worst nuclear power plant accident occurred April 26, 1986 when an explosion and fire split open the Chernobyl atomic reactor in the Ukraine and sent clouds of radioactive debris into the atmosphere. Radioactive fallout carried by strong winds circled the globe.

The Chernobyl explosion left at least 27 nearby cities and villages too contaminated for people to live in for the foreseeable future. The accident left 31 dead in the Soviet Union; 135,000 had to be evacuated.

Two years after the disaster, beef and dairy products in Ireland, Denmark, Great Britain, Italy and the Netherlands were still contaminated. Sweden's annual reindeer kill (about 100,000 animals) had to be disposed of since the radiation levels made the meat too dangerous to eat. Expectant mothers in the most contaminated areas were warned to follow strict diets. Vast herds of contaminated sheep remain isolated in Britain. In some areas fish, fruits and vegetables could not be eaten. Many European governments returned Turkish tea and hazelnuts when they were found to be contaminated. Countries in Latin America returned or refused to accept some European foods.

Our study compares the amount of radioactive fallout expected ten days after a given nuclear strike scenario with the estimated total radioactive release at Chernobyl as calculated ten days after the accident. This yardstick is very crude, since nuclear weapons fallout has a faster decay rate than that of reactor fallout. Nevertheless, such a yardstick may at least dramatize the unprecedented catastrophe threatened by radioactive weapons.

In order to determine the amount of radioactive fallout a nation might inflict upon the world, and itself, we assume in each of the following estimates that only a nation's own nuclear weapons are used. In order to dramatize this self-destructive

21

aspect of nuclear weapons, we assume that there is no retaliation; such retaliation, of course, could mean far greater destruction. We do not attempt to estimate the many millions of casualties that could result promptly from heat and blast effects, nor the even larger number of deaths which could result from agricultural and social disruption on a national or even global scale. Relying only on unclassified data and press reports, we have made reasonable estimates as to the number and size of weapons if actual numbers are not available.

United States — Trident II

In order to show how environmentally destructive even a single weapons system might be, let us consider what could happen if just one U.S. submarine, acting on its own, mistakenly launched its Trident II nuclear missiles against the (former) Soviet Union, and no other nuclear weapons were exploded. Such a mistake could produce an amount of radioactive fallout, much of it globe-circling, equivalent to from 107 to 337 Chernobyls depending on number and size of warheads.

The U.S. plans to spend $85 billion to build about 20 Trident submarines, associated weapons equipment and shore support. These subs would carry enough missiles to produce fallout equal to 1,780 to 6,740 Chernobyls. Have military planners assessed the number of cancer deaths and birth defects that might be expected from global fallout on the U.S. as well as on its allies and neutral countries?

Great Britain — Trident II

British ship and submarine crews, like those of the United States, can use nuclear weapons on their own under certain circumstances. Britain plans to build four Trident submarines during the 1990s to replace its current four older subs. If one of these Trident submarines were to intentionally or mistakenly launch its missiles, this could create radioactive fallout equal to 71 to 225 Chernobyls. In the wake of the Chernobyl accident, are British authorities assessing how lethal this fallout could be?

The Former Soviet Union and the U.S.

If either the United States or Russia accidentally or intentionally launched a major first strike (5000 Megatons) against the other side and no other nation used any nuclear weapons, the attacker's weapons could produce worldwide radioactive fallout equivalent to 18,500 Chernobyl nuclear power plant accidents.

Much of this fallout would circulate globally, with especially heavy concentrations at latitudes 30-50 degrees North and 50-70 degrees South. Thus a large part of the global fallout would be deposited within the borders of the nation which launched the strike and also within the border of its allies as well as neutral countries.

Israel

By focusing our attention on nations with large nuclear arsenals, we tend to overlook what might happen if smaller arsenals were used. For example, let us consider what would happen if Israel used its estimated 100 or more nuclear weapons. Let us assume that in order to reduce radioactive "backlash" to their own territory, the Israeli weapons-makers have built comparatively "clean" hydrogen bombs with an average yield of 100 kilotons each—20 percent fission, 80 percent fusion. If Israel used its weapons and no other nation used nuclear weapons, this could release radioactive fallout equivalent to about 15 Chernobyls. Since the targets could be nearby, much of the intense fallout in the first 24 to 48 hours after a nuclear strike could still land in Israel itself. Airbursts over the targets would minimize this hazard of intense local fallout, but could double the amount of global fallout.

France

France has more than 400 nuclear weapons on its aircraft, land-based missiles and submarines. If France were to use only a small part of this force, say its 18 intermediate-range ballistic missiles, this could produce total radioactivity equivalent to 67 Chernobyls. France plans to replace its intermediate-range missiles with more powerful ones. Has anyone studied the possible impact upon France and its allies from global fallout?

China

If China were to use its 300 nuclear weapons and no other nations used nuclear weapons, this alone could release radioactivity equivalent to 1900 Chernobyls.

Other Hazards

Nuclear weapons can create serious radioactive hazards even if they are never fired. For example, if the conventional explosive trigger of a nuclear weapon were detonated by accident, it could scatter highly toxic plutonium over a wide area. A typical nuclear weapon contains more than four pounds of plutonium. Inhaling or ingesting as little as 2/10,000 of an ounce of plutonium dust can cause fatal illness.

Like the Chernobyl cloud, fallout from a nuclear war would be no respecter of national or ideological boundaries. Therefore, all nations and people have a vital interest in the control and reduction of nuclear arsenals. Indeed, official bodies in non-nuclear states such as the Swedish Academy of Sciences have sponsored important studies on the consequences of nuclear war. Can the governments of the nuclear states afford to show less concern?

The nuclear states may find that all of their options for using nuclear weapons could lead straight to self-destruction. Such states could then join with other nations in budgeting their funds for the improvement rather than the destruction of the global environment.

THE WAR ON NATURE:
AN OVERVIEW

TOXIC TRAVELS

Seth Shulman

Seth Shulman is a freelance writer and correspondent for Nature. *His book, published in the spring of 1992, is titled* The Threat at Home: Confronting the Toxic Legacy of the U.S. Military.

Points to Consider:

1. How has the military damaged the environment at the Jefferson Proving Grounds?

2. What will it cost to clean up the military's toxic waste legacy? Give at least two specific examples.

3. Why is the military described as a vast industrial complex?

4. Why did the Rocky Mountain Arsenal haunt the author so?

Seth Shulman, "Toxic Travels," **Nuclear Times,** Autumn 1990.

The toxic legacy left by our nation's military in-frastructure constitutes the largest and most serious environmental threat this country has ever faced.

Heading south on Route 421, somewhere past the tiny Hoosier town of Versailles (pronounced Ver-SAYLes), lies the U.S. Army's Jefferson Proving Ground. This southeastern corner of Indiana is beautiful American heartland, with miles of rolling pastures and cornfields, situated near the Ohio and Kentucky borders. But here, as in many locales around the nation, an ominous, toxic legacy has surfaced to threaten us all. The question, still very much unanswered, is whether we can successfully meet the daunting challenge this legacy presents.

Enclosed behind an uninviting 48-mile perimeter is one hundred square miles of Indiana that will likely remain closed off forever: an area more than four times larger than Manhattan permanently isolated from human contact like a quarantined victim with a contagious and terminal disease. The disease afflicting this vast chunk of Indiana countryside is military toxic contamination.

For its size and quality of contamination, Jefferson Proving Ground presents an unparalleled environmental dilemma, but the facility's problems are not unique. Hidden from public view and largely unfettered by environmental regulations, the U.S. armed forces have left a shocking and varied legacy of contamination at virtually every military installation in the country and at hundreds more bases around the world.

Because of its vast size, the U.S. military continues to rank among the world's largest generators of hazardous wastes, producing nearly a ton of toxic pollutants every minute. Despite recently begun "waste minimization" programs, the military continues to dump large quantities of deadly chemicals improperly—with little oversight or public accountability.

Today, hazardous wastes are suspected of contaminating more than 20,000 sites on land currently or formerly owned by the U.S. military. At these locations, million of tons of military toxins have fouled many hundreds—if not thousands—of square miles of soil and polluted air and groundwater in communities across the country, and undoubtedly at hundreds of overseas bases as well.

The obstacles to environmental restoration are formidable, and not the least of them is money. Between the wastes from the Energy Department's nuclear production facilities and those of

Cartoon by Michael Powers, Nonviolent Activist, **War Resister's League**

the Pentagon's bases, the job is now expected to cost several hundreds of billions of dollars.

A Shocking Legacy

After more than a year of research, I have caught only a glimpse of the total picture. But my toxic travels lead me to believe, as some in the Pentagon and the Department of Energy are now beginning to acknowledge, that the toxic legacy left by our nation's military infrastructure constitutes the largest and most serious environmental threat this country has ever faced.

At Jefferson Proving Ground, the army has tested huge quantities of conventional munitions since World War II. After fifty years of discharging some 23 million rounds of ordnance, the army has littered the vast site with more than 1.5 million unexploded bombs, mines, and artillery shells. Some of the ordnance, buried as deep as thirty feet below the surface, are white phosphorus shells that officials at the facility say are certain to ignite if ever exhumed. JPG, as the facility is known locally, is also home to low-level radioactive contamination, toxic sludge, and pesticide residues.

JPG tests 85 percent of the army's conventional munitions and currently fires some 80,000 rounds annually—nearly one every minute when the facility is operatin .JPG's environmental problems were placed in clear relief recently when the base became one of 86 military installations around the country

Congress ordered shut in the first round of base closures.

Since the base closure order became final in 1989, a new picture has emerged. Now, it seems, a complete cleanup may be too dangerous and destructive to conduct, not to mention prohibitively expensive. To remove all the bombs, most of JPG's hundred square miles of wooded and bombed-out land would have to be stripped down to a level thirty feet below the current surface, using special armored bulldozers.

In many ways, JPG is emblematic of the nation's military toxic mess. The military's roster of toxic sites has tripled since 1986. Even between the two most recent annual reports—for 1988 and 1989—the military discovered more than 5,000 new suspected sites at its installations, the bulk of them at army bases.

Despite the military's current rhetoric, their cleanup record to date is dismal, displaying an unprecedented pattern of environmental abuse and neglect that spans generations, as the following examples attest.

Lakehurst

At New Jersey's Lakehurst Naval Air Station, a 1983 navy report documented that 3.2 million gallons of cancer-causing aviation fuel and other toxic substances had polluted an aquifer that provides tap water for most of southern New Jersey. Three different tests indicated levels of toxic substances as high as 10,000 times above levels the state considers safe.

The navy's report was not made public until two years later—after the local press had independently investigated the contamination. Lakehurst has yet to be cleaned up.

Cornhusker

At the Cornhusker Army Ammunition Plant in Nebraska several years ago, an army spokesperson defended the base's decision not to notify nearby residents that their drinking wells were contaminated with dangerous levels of explosive compounds. "We didn't want to get them overly anxious," he said.

McClellan

With a total of 167 separate toxic wastes sites, McClellan Air Force Base in Sacramento, California, ranks as one of the most contaminated installations in the country. Much of the waste comes from solvents the air force routinely uses to spray down its planes. For decades, they drained the chemicals directly into the ground. But the military also admits to the presence of high levels of degreasers, PCB's, acids, low-level radioactive waste, and other contaminants.

Hanford

At the Energy Department's notorious Hanford Nuclear Reservation in the southeastern corner of Washington State, one million gallons of high-level liquid radioactive waste have reportedly leaked from underground tanks. And more than a half-million curies of radioactive materials have been released into the atmosphere, exposing at least 13,000 residents to dangerous levels of radiation. The Energy Department kept the sanctioned airborne emissions secret for forty years. The information was discovered only after local residents fought for the release of over 19,000 pages of government documents from the facility.

The cost to clean up Hanford's radioactive and mixed toxic wastes is now estimated at a staggering $57 billion, more than four times greater than the Energy Department's entire budget for fiscal year 1989. In addition, much of this effort still remains technically unresolved; planners have no idea how to clean up the leaking and potentially explosive tank farm. Even if such solutions were found, no permanent dump site yet exists for this high-level nuclear waste.

Standard Operating Procedures

The first key to understanding the military toxic waste legacy is an appreciation of the Defense Department's huge scale. The bloated size of the military is often expressed in terms of its $300 billion yearly budget or its two million enlisted personnel, but the scale can be illustrated in other ways as well.

The emerging story of military toxic waste reveals a picture of the Defense Department as a vast industrial enterprise. It purchased, for example, over 200 billion barrels of oil for military use in 1989 — enough, by one estimate, to run the nation's entire public transit system for 22 years. The armed forces are believed to have some 40,000 underground storage tanks to hold oil and other chemicals, and many of these tanks are known to be leaking.

The bulk of the military's toxic wastes originates from decades of standard daily operating procedures. Many practices have been changed to meet environmental regulations; many others continue unaltered since World War II, and even earlier. These standard procedures are used for developing, producing, testing, storing, and modernizing the military's conventional weapons as well as its nuclear and chemical arsenals.

But Jefferson Proving Ground gives more than ample illustration that the environmental problems associated with conventional munitions do not come only from their production. In addition to the military's weaponry, however, huge quantities of waste are generated through the construction and maintenance of military equipment. Each branch of the armed forces maintains a huge fleet of vehicles, tanks, planes, or ships, all of which require the routine use of hazardous materials.

Needless to say, the emerging picture is troubling. "What good does it do to protect ourselves from the Soviets," an exasperated Senator John Glenn (D-OH) asked his colleagues, "if we poison ourselves in the process?"

Spoils of War

Travels for work and pleasure have taken me to many spots around the world where the haunting and palpable memory of war lingers. Few experiences, however, prepared me for the more hidden consequence of war that I was to witness at the Army's Rocky Mountain Arsenal on the outskirts of Denver. To me, its lasting image is more potent than books full of data about military contamination.

For decades, some of the most deadly chemicals known have

31

been dumped on the arsenal's land, including lethal by-products of nerve and mustard gas production. The contaminants have conspired to lend the facility its dubious distinction as home to the most toxic square mile on earth.

The row of dump sites, each the size of a small lake, visibly marks the arsenal's history. At the far end, hundreds of yards across the road, lay the unforgettable sight of the most recent dump—Basin F. Because it had been lined with asphalt, unlike its neighbors, the 93-acre Basin F still displayed its noxious contents: nearly 9 million gallons of aquamarine-colored toxic sludge. Only at the sight of Basin F did the horror of the other now-empty dump sites come into perspective. Only then, before this phosphorescent toxic lake, nestled beneath the Rocky Mountains, could I begin to fathom the arsenal's poisonous underground plume, nearly the size of the entire 26-square-mile facility, migrating inexorably northwest, contaminating groundwater on the edge of Denver and threatening the health of neighboring residents.

Since my visit, the sludge from Basin F has been sucked into large holding tanks, and the residue has now been piled into a huge mound. But the memory of the deadly chemical lake—and its five ghostly predecessors—will always remain for me as a haunting symbol of war's insidious toxic legacy, one as powerful as any I've seen.

Today, only a skeleton crew remains at Rocky Mountain Arsenal. The facility's mission—to manufacture munitions to protect us against a foreign enemy—has now ceased. In the name of defending our national security, however, Rocky Mountain Arsenal has created a different kind of threat to our land, water, air, and to our health.

Sadly, the chances are good that you, too, live near a military base. Your community's air and drinking water may also be threatened. Like me, you may justifiably feel anger and outrage about the military's sorry stewardship of your neck of the woods. My outrage remains, but my own struggle against military contamination has taught me that, with increased public scrutiny and a shrinking military infrastructure, we may be able to change current military practices and ensure a careful and thorough cleanup process.

RECOGNIZING AUTHOR'S POINT OF VIEW

This activity may be used as an individualized study guide for students in libraries and resource centers or as a discussion catalyst in small group and classroom discussions.

Guidelines

Good readers make clear distinctions between descriptive articles that relate factual information and articles that express a point of view.

A. Write down two statements from each of the readings in Chapter 1: one factual statement and one statement that presents a point of view.

B. Read through the following statements and indicate with an (**O**) those statements that express an opinion or point of view, and use an (**F**) for any factual statements. Use a (**?**) for statements you are not sure of.

___ 1. Even in "peacetime", the military contributes to environmental degradation.

___ 2. Cleaning up after the military will cost millions worldwide.

___ 3. The military sector has long considered itself beyond environmental rules and regulations.

___ 4. In order to ensure international ecological security, it is necessary to adopt binding principles and norms on the conduct of nations.

___ 5. More than 15,000 objects have been discarded in orbit around the earth.

___ 6. Every would-be world leader should be shot into space for one week before assuming office.

___ 7. One Trident submarine could unleash the fallout equivalent of 377 Chernobyls.

___ 8. All nations and all people have a vital interest in the prevention of nuclear war.

___ 9. Militarism and its damage to the environment is an unfortunate reality in preserving national security.

___ 10. Halting the arms race will encourage the establishment of peace and security, thus promoting sound ecological policies.

___ 11. The time required for decontaminating sites polluted by military toxic and radioactive wastes may be measured in decades and generations.

___ 12. The destruction of our global natural support systems is a steep price to pay for a militarized national "security".

CHAPTER 2

GLOBAL MILITARY POLLUTION:
Four Case Studies

5   GLOBAL MILITARY POLLUTION

THE VIETNAM WAR THAT NEVER ENDED

John Schuchardt

John Schuchardt is a Massachusetts attorney and U.S. Marine Corps veteran. He is also a member of the Plowshares Eight, a group of antiwar activists.

Points to Consider:

1. What is the extent of environmental damage to Vietnam?

2. How does the damage from the war continue today?

3. What is Agent Orange, and how much was used in Vietnam?

4. How many Vietnamese have been affected by exposure to chemical defoliants?

John Schuchardt, "Ecocide in Vietnam: A War That Never Ended," **Toward Freedom,** 1990.

The 18 million gallons of herbicides sprayed on the southern part of Vietnam long ago entered the food chain.

Ecocide is a term coined in an attempt to describe the effects of the U.S. war against the environment of Vietnam. My impression of the continuing ecological devastation of Vietnam is consistent with data compiled by Dr. Arthur Westing of the Stockholm International Peace Research Institute.

A Land of Craters

Westing estimates that the southern half of Vietnam contains 21 million large-sized bomb craters. They were gouged out of the earth by the explosives dropped on Vietnam from U.S. B-52 bombers during a decade-long assault that was most intensive in the South.

These holes are not very visible at ground-level, unless one is standing right at the edge of a typical crater measuring 15 to 25 feet in depth. From the air, however, the land resembles a vegetated moonscape. Moreover, the fields and forests of southern Vietnam are still seeded with millions of unexploded mines, bombs and artillery shells. Many of them are quite small and difficult to detect. M-79 grenades, for example, are about the size of an egg, while gravel and dragon-tooth mines are no bigger than books of matches. Countless numbers of these devices were scattered over the countryside as part of a program known as "area denial".

The mines and cluster bombs were often designed to detonate only upon a second impact—or, when stepped on by a farmer years after the initial bombing. In fact, the U.S. is continuing to add 3000 Vietnamese civilian victims each year to the casualties it has been causing in the country since 1954.

Deliberate Ecocide

It is also fairly well known that the U.S. deliberately destroyed much of Vietnam's flora through indiscriminate use of defoliants. But it was not until I visited the Tu Du Obstetrical and Gynecological Hospital in Ho Chi Minh City (formerly Saigon) that I truly understood the meaning of the word, "defoliant".

The 18 million gallons of herbicides sprayed on the southern part of Vietnam long ago entered the food chain. Poisons such as Agent Orange contain dioxin, a non-biodegradable substance and a particularly potent carcinogen.

37

CHINA

NORTH VIETNAM

Ha Giang

Song Co River

Hanoi

Haiphong

Dien Bien Phu

LAOS

Samneua

Nam Dinh

Thanh Hoa

Gulf of Tonkin

HAINAN

Luang Prabang

Long Tieng

Vinh

Vientiane

Nong Khai

Udorne

Dong Hoi

DMZ

Hue

Da Nang

My Lai

THAILAND

Saravane

Paksee

Bangkok

Khong

Kontu

Pleiku

Angkor Wat

Stung Treng

Siem Reap

Battambang

CAMBODIA

SOUTH VIETNAM

Kompang Chnang

Nha Trang

Cam Ranh

Phnom Penh

Kompong Som

Prey Veng

Bien Hoa

Gulf of Siam

Con song

INDOCHINA

Craterization

Defoliation

0   100   200

1970

Daily World

Genetic Destruction

The Vietnamese estimate that 1.3 million of their citizens have developed disorders attributable to exposure to defoliants. In addition to cancer, these afflictions include degenerative diseases of the kidney, liver, heart and central nervous system. Insomnia, amnesia and chloracne are also common among these victims.

Defoliants have likewise caused enormous genetic destruction in Vietnam, resulting in thousands of stillbirths, molar pregnancies, gross fetal malformations and co-joined babies. The doctors and teaching staff at Tu Du Hospital, where we witnessed some of these cases, conveyed an urgent plea for normal relations between the U.S. and Vietnam. They explained that research and technical assistance from the industrialized nations is needed in order to help stem the genetic destruction.

In an attempt to educate North Americans about the ongoing impact of the U.S. war on Indochina, I have been presenting a slide show entitled, " Vietnam: Reconciliation and Healing". Slides are an effective means of communication, I believe, because images can bridge the split between thinking and seeing, between analysis and feeling.

6   GLOBAL MILITARY POLLUTION

NUCLEAR THREAT
IN THE PACIFIC

David Robie and Andre Carothers

David Robie is a New Zealand based writer specializing in Pacific affairs.
Andre Carothers is the staff editor of Greenpeace Magazine.

Points to Consider:

1. Why are France and the U.S. interested in continued control over regions of the South Pacific?

2. How many nuclear "tests" have the French military conducted?

3. What effects have they had on the islanders' health and environment?

David Robie and Andre Carothers, "Rising Storm in the Pacific," **Greenpeace,** January/February 1989.

After exploding 44 bombs above ground, France finally agreed to go underground in 1975.

The Pacific Ocean covers nearly a third of the world's surface and is home to more than five million indigenous islanders. And for 400 years, these people have lived under the flag, nominally and otherwise, of various Western nations. At first, Western interests in the scattered islands were largely economic. At the time of Captain James Cook's "discovery" of New Caledonia in 1774, noted one historian, voyages of exploration in the region were "motivated as much by scientific curiosity as by the lure of the gain."

But the promise of riches in the scattered islands proved illusory. As a result, colonialism tended to be superficial. Thus, while Africa, Asia and the Caribbean were embroiled in intense and often violent liberation struggles after World War II, the Pacific enjoyed tranquil or even stagnant relations with its metropolitan overlords.

It was not until the late 1960s that a fresh political geography began to emerge in the South Pacific. In a decolonization process that was remarkably peaceful, name changes and new nations altered maps, as the last outposts of the European empires were cast adrift.

Today, France and the United States are the only other countries retaining clear and complete control over regions of the South Pacific. The concrete manifestation of this presence is nuclear; both nations have built extensive nuclear-related facilities in the region, and France maintains a nuclear testing program that has exploded nearly 150 bombs in the fragile basalt beneath Polynesia's Moruroa atoll. As a result, the aspirations for independence of the people of the Pacific have become paired with their desire to be free of nuclear weapons, giving rise to the catch-all name the loose Pacific coalition has taken on: the Nuclear-Free and Independent Pacific movement.

France Tests the World's Patience

In 1966, bowing to the inevitable end of its colonial control in Algeria, the French government shifted its nuclear test site from the Sahara to the South Pacific island of Moruroa, in French Polynesia. Rejecting the ban on atmospheric tests negotiated by the Soviet Union and the United States (such an agreement, they argued, would "hamper French independence"), France persisted in detonating nuclear bombs tethered to balloons above the atolls.

Illustration by Ron Swanson

After exploding 44 bombs above ground, France finally agreed to go underground in 1975, but not before it had been dragged to the World Court by Australia and New Zealand, weathered a regional boycott of French goods, and had photographs of French paratroopers beating up the skipper of the Greenpeace protest ship Vega splashed across the world's newspapers.

"It has become increasingly apparent," says Andrew Burrows of the Natural Resources Defense Council, "that an atoll is possibly the worst place to conduct underground tests." Rocked by 100 nuclear explosions in a dozen years, the fragile basalt beneath the atoll has crumbled. In addition to sporting an enormous underwater crack, the whole atoll has subsided nearly five feet. A team headed by Jacques Cousteau released a report on a 1987 visit to the atoll, including underwater film showing "spectacular fissures" resulting from the nuclear tests.

In the course of the testing program, France reportedly turned the north end of the atoll into a radioactive waste dump and spread 10-20 kilograms of plutonium-impregnated tar across one section of Moruroa as part of a security exercise for nuclear accidents. The cyclone of March 11 and 12, 1981, probably swept the entire mess into the sea, but France won't confirm this.

France's reluctance to allow independent examination of its Moruroa program is legendary. Following the 1981 cyclone, worried French technicians at the site leaked to the world press information on the waste dump and the training exercises, earning a reprimand from their employers for "breaches of security". The two "independent" investigations that were conducted before 1988, the Tazief Mission in 1982 and the Atkinson Report a year later, spent a total of about two weeks on the atoll and had limited access to the test site. The Tazieff Mission was allowed to witness a subkiloton test, one of the smallest ever conducted at the site, and the Atkinson team was restricted to examining rock and coral samples supplied by the test site managers.

Health Threats

Health studies of the inhabitants of neighboring islands are also hampered by France's refusal to cooperate. The involvement of France's National Radiation Laboratory is prohibited, and in 1966, when the test site was established, the French government stopped releasing health statistics. According to statistics compiled by the South Pacific Commission, cancer deaths in Polynesia average 40 a year, up from five in 1959, zero in 1960 and eight in 1962. The influence of better health reporting and increased lifespans on these cancer statistics is difficult to measure.

France's most callous act was the sinking of the Rainbow Warrior. In July 1985, French secret service agents attached underwater mines to the hull of the Greenpeace flagship while it was docked in New Zealand, killing photographer Fernando Pereira. The agents were acting under orders from Paris to "anticipate" the ship before it joined in a protest against French nuclear tests. The capture of two French agents in New Zealand eventually unraveled a cover-up that led up to the office of French President Francois Mitterrand. In the ensuing scandal, France's defense minister, Charles Hernu, resigned, as did the head of the French secret service.

Nuclear-free and Independent

As old guard leaders in the Pacific islands either die or lose authority at the ballot box, younger, more radical leaders are emerging to take their place. Among national leaders with a more progressive outlook are Vanuatu's Father Walter Lini, President Ieremia Tabai of Kiribati and Fiji's Bavadra. Others in nationalist movements include Jean-Marie Tjibaou of New Caledonia and Tahiti's Oscar Temaru and Jacqui Drollet.

The newcomers have given a dynamic and crusading voice to the region's unspoken wishes. "The great ocean surrounding us carries the seeds of life. We must ensure that they don't become the seeds of death," said Tjibaou in a moving speech before the 1985 signing of the Rarotonga Treaty, "A nuclear-free Pacific is our responsibility, and we must face the issues to live and protect our lives."

7   GLOBAL MILITARY POLLUTION

DEVASTATION IN CENTRAL AMERICA

Environmental Project on Central America

EPOCA, the Environmental Project on Central America, has been discontinued. The Material Aid Group of EPOCA called Appropriate Technology Working Group (A.T. WORK) continues to maintain the files and library of EPOCA at 300 Broadway, Suite 28, San Francisco, CA 94133-3312. Phone number 415-788-3666.

Points to Consider:

1. Who was responsible for the military assault on the people and ecology of Central America in the 1980s?

2. Describe "scorched earth" as practiced in Guatemala.

3. How did militarism affect the forests and wildlife of El Salvador?

EPOCA, "Militarization: The Environmental Impact," **Earth Island Journal,** Fall 1986.

*The "low-intensity warfare" strategy employed by the
United States government has taken a toll on the
people and environment of the region.*

Central America's ecological diversity rivals any in the world.
From the iridescent blue flashes of the Morpho butterfly in a
Panamanian rain forest, to the legendary quetzal whose rare
shadow burns into the mist of Guatemala's highlands, the
region's natural phenomena have dazzled generations of
naturalists. For eons this crumpled isthmus has served as a
bridge and transition zone for tropical and temperate lifeforms
from North and South America. Medical and agricultural
treasures lie hidden in the more than 1,000 species of plants
found nowhere else in the world.

Years of Degradation

The "low-intensity warfare" strategy employed by the United
States government has taken a toll on the people and
environment of the region. Bombs burn bodies and scorch the
land in El Salvador. Military roads, bases and air strips
criss-cross the Honduran and Guatemalan wilderness. Among
those being targeted in Nicaragua by the U.S.-backed contras
were professional environmentalists. Hundreds of thousands of
Central American refugees fled their homes in search of safe
haven, often settling in remote forested areas.

Militarization also shifts people's attention from public welfare
to warfare. Funds that could support social and environmental
programs are diverted into a widening stream of military
hardware. The environmental consciousness that has emerged in
the last decade is overwhelmed by a military mentality.

Bombarding El Salvador

The U.S.-backed Salvadoran government and the
counter-insurgency policy being employed by the Salvadoran
Armed Forces attempted to separate the guerrillas from their
support systems—the rural population and the natural
environment. The air war in El Salvador—the bombing of
villages, crop lands and forests—reached levels unprecedented
in Central America. Fragmentation and incendiary bombs may
have irrevocably damaged the already ecologically overburdened
Salvadoran countryside, killing and wounding innocent people in
the process.

In the northern province of Chalatenango, government

bombing raids have destroyed forests and fields, making cultivation impossible. According to a priest in the area, "these bombs leave craters 15 feet deep and sever trees too thick to encircle with one's arms." In San Vicente province, the bombing of agricultural fields has created an acute food crisis, depriving peasants of corn, beans and rice. El Salvador's Air Force dropped more than 3000 tons of U.S.-made bombs on its own civilian population, killing at least 2,000 people.

Some of the weaponry used in Vietnam has resurfaced in El Salvador. For example, "daisy cutters" have found a niche in the Salvadoran government's arsenal. These anti-personnel bombs explode just above ground level, blasting a horizontal force that clears the area not only of human beings, but shaves the forest floor, leaving a stubble of splintered trunks. The Salvadoran landscape also bears the scars of napalm and white phosphorus bombs. In Cuscatlan, sugar cane plantations and coffee crops have suffered from 200-pound bombs, as well as from napalm and white phosphorus.

Wildlife populations also suffer. Troop movements, bombings and forest fires undoubtedly affect deer, small cats, howler monkeys, sloths and iguanas. Moreover, as Central America is on the flyway of many migrating bird species, the war threatened the wintering habitats of North American songbirds—tanagers, warblers, kingbirds and vireos.

Guatemala: Landscape of Violence

Thirty percent of the 100,000 deaths credited to the Guatemalan army have occurred in the 1980s. Focusing in the highland provinces of El Quiche, San Marcos, Huehuetenango and Alta Verapaz, the army used tactics of ecological warfare. Analyzing this phenomenon, one western observer noted that "it was done under the guise of making the world safe for democracy, and getting rid of communism. But everybody knows what was happening. They were going and taking out their own people. Scorched earth."

The Guatemalan armed forces have destroyed forests, fields, livestock and at least 440 villages, displacing more than one million Guatemalans from their land. When explaining the army's tactics to a journalist in 1983, an Indian resident noted that "half the forests in Chichicastenango have been destroyed by fire so far this year, depriving us of wood for cooking and destroying the natural resources which could help us to survive in the future." Recent reports from the area indicate that so much of the forest has been destroyed that the micro-climate is changing and temperatures are increasing.

Guatemala's scorched earth policy is having deep and long-lasting effects on the country's Indians, who make up two thirds of its 8.3 million inhabitants. Counter-insurgency targets corn cultivation. By destroying Indian crops, the army breaks age-old bonds between community and environment, decimates a valuable genetic reserve, and causes soil erosion that renders large areas virtually useless for years.

The Guatemalan army also built a series of roads into remote areas to combat the guerilla insurgency. Its counter-insurgency tactics destroyed forests along roads and near villages as well. Forests have suffered a pincer attack from both the guerrillas, who felled trees to roadblock army convoys, and the army, trying to avoid such ambushes. Army bulldozers, herbicides, machetes and flamethrowers quickly denude the lush rain forest and highland terrain.

Peace Through Parks

An innovative concept known as "Peace Through Parks" has recently emerged in Central America. Parks that span borders could play a crucial role in easing tensions on the isthmus, while protecting a number of threatened tropical ecosystems and the many species of wildlife now in danger of extinction in the region. Inter-country biosphere reserves could significantly contribute to demilitarization and sustainable economic development in wilderness areas. Peace Through Parks will help to change the focus from rhetoric and warfare to peaceful cooperation between nations.

Peace Through Parks alone cannot solve the military and environmental crises in Central America and, in fact, cannot be successfully implemented unless it is accompanied by a more general move towards demilitarization in the region.

As environmentalists who understand that nature's forces cannot compete with armed forces, we need to work to preserve Central America's diverse natural beauty before it is too late.

8   GLOBAL MILITARY POLLUTION

ECOCIDE IN THE GULF

Brian Tokar

Brian Tokar is a freelance writer who produced this article on the environmental effects of the Persian Gulf War.

Points to Consider:

1. How has oil become such an ecological threat?

2. Will the Persian Gulf recover from the ecological damages of the war? Explain.

3. What effects have allied bombing had on the environment and civilian populations in Iraq?

4. How does military fuel consumption illustrate the main theme of this reading?

Brian Tokar, "War Is Ecocide," **Z Magazine,** April, 1991.

Even though the Persian Gulf War has ended, the ecological consequences of the war against Iraq will linger for generations.

War has always been an ecological disaster. From the defoliated jungles of Vietnam to the scorched and cratered hillsides of El Salvador, modern warfare has proved brutally unrelenting in its ecological, as well as human consequences. The pursuit of military victory in the age of mass destruction weapons means, literally, not leaving a single stone unturned.

The environmental consequences of war may be easy for some people to overlook, in light of devastating losses of human life and the social and political hysteria that are rampant during wartime. But the consequences for the earth are invariably the most lasting, and they continue to plague people and their land long after hostilities have ended. Every week hospitals in Vietnam still report several new victims of leftover land mines from that ecocidal war. Much of the country's land remains unable to sustain life, as the effects of chemical defoliants dropped by United States forces continue to linger.

In the desert, it can be much worse. The scars and stirred-up dust from World War II tank battles still mar the sands of Libya, and sparse soils there may not fully recover for hundreds, or even thousands, of years. Even though the Persian Gulf War has ended, the ecological consequences of the war against Iraq will linger for generations.

A Fragile Land

When most people think of the Middle East, they imagine a barren desert land, completely hostile to life except for a few isolated oases. This picture is partly true, of course, especially for the Arabian peninsula, whose deserts and arid steppes stretch on for hundreds of miles.

But the Middle East is not all desert. The southwest of what is now Saudi Arabia has been known since ancient times for its date palms and other important fruits. The mountains of Oman, past the mouth of the Persian Gulf, support a rich agriculture. The Tigris and Euphrates Rivers of Iraq carry water down from the high peaks of Syria and Turkey to form the fertile valleys from which Western civilization first began to emerge. The Mediterranean coasts and interior lands of Lebanon and Syria support olives, grapes, and other native plants, as well as a variety of subtropical fruits.

Sketch by Ron Swanson

The native fauna of the Middle East include panthers, leopards, hyenas, wolves, foxes, jackals, hares and lizards. Birds of prey such as eagles and falcons are common, and the Persian Gulf itself is known for its native dolphins, sea turtles, and a rare sea cow known as the dugong, which is a close relative of the manatee.

Oil as a Weapon

The discovery of oil in the 1920s brought unprecedented new stresses to fragile Gulf ecosystems. Today, there are 25 major oil terminals serving over 20,000 tankers in a normal year. There are more than 800 active offshore wells, and pipelines literally line the bottom of the sea in some places. Oil spills totalling 150,000 metric tons a year are considered "normal". In 1983, Iraq blew up Iran's Nowruz drilling platform, releasing more than 20 million gallons of oil into the Gulf over an eight month period. The bombing of Iran's Sirri Island terminal in 1986 killed over 500 dolphins and a whale, among other sea animals. The United States also had its turn during the Iran-Iraq war: in 1987, the U.S. Navy shelled two oil platforms off the Iranian coast, and continued shelling long after the platforms were completely destroyed. During that war, the United Nations Environment Program described the Persian Gulf as "one of the most fragile and endangered ecosystems" in the world.

Not only was oil a leading cause of the war; it rapidly became one of its major weapons. The proximity of the fighting to the Gulf's vast oil terminals presented a serious ecological threat, but the targeting of oil facilities, the burning of Kuwaiti wells, and the deliberate release of oil into the Gulf's waters compounded this threat.

As early as January 22, 1991, Iraq began setting fire to oil facilities in Kuwait, destroying at least two major storage tanks. Two days later, the U.S. bombed an Iraqi oil tanker said to be engaged in battle reconnaissance. On the 26th, Iraq began to pump oil from Kuwait's Sea Island Terminal, ten miles offshore, creating one of the largest oil spills in world history.

It is unclear whether the Gulf itself, with its rich and diverse offshore ecosystems, will ever fully recover. Unlike the coast of Alaska, where turbulent waves replenish sea water in a matter of months, the Persian Gulf is isolated from the open sea by the narrow Strait of Hormuz. Water is replenished only over periods of many years or even decades. The Gulf is also a shallow sea, with an average depth of only 110 feet, so even oil that sinks to the bottom will continue to be damaging to marine life. The tidal zone extends for as much as a mile offshore, and oil is trapped during low tide in biologically important mud flats. Mid-winter was also the spawning season for the Gulf's shrimp populations, which are undoubtedly severely threatened by the oil slick.

An even more serious threat was posed by the Iraqi mining of Kuwait's 750 oil wells. The burning of oil wells began in the second week of the war. Massive amounts of soot from burning oil have fallen back to earth in the form of "black rain", which forms when soot particles begin to saturate rain clouds. Black rain was reported as far as 150 miles from the battlefields, high in the mountains of Iran, even before the massive oil fires.

Ecocidal Nightmares

The U.S. military would have us believe that their precise "surgical strikes" did not kill many Iraqi civilians. They also claimed to be destroying Iraq's capability to launch chemical and even nuclear weapons without damaging non-military facilities. The myth of no civilian casualties was put to rest when hundreds were killed in one attack on a reinforced bunker in Baghdad in February. The consequences of attacks on chemical and nuclear installations can still only be speculated upon, but,

THE CHILDREN SUFFER

Children have been among the war's greatest victims. Death rates for children under the age of five have risen 390 percent over pre-war levels. An estimated 900,000 children — nearly a third of Iraq's youth — now suffer from malnutrition. Malaria and polio also pose a threat to Iraq's children.

The psychological scars are equally devastating. Nearly two-thirds of all the children interviewed said they believed they would never live to become adults.

Ross Mirkarimi, "After the War, Iraq Still Bleeds," **Earth Island Journal,** Winter 1992

given the proximity of Iraqi weapons plants to population centers, such attacks could be far more damaging than the use of these weapons on the battlefield. The full extent of the damage, and thus the consequences, will not be known for some time.

Iraqi stockpiles of chemical, as well as biological, warfare agents have been estimated in the thousands of tons. The largest storage and production facility for these agents in the region was about 25 miles southwest of the Shi'ite holy city of Samarra, northwest of Baghdad. Several military installations, power plants, chemical facilities and airfields are in the region, so it was undoubtedly a target for Allied air strikes. The bombing may well have created uninhabitable chemical zones, as well as contaminating rivers and canals that feed the Tigris River upstream from Baghdad.

The nuclear threat to the region is more clearly the result of U.S. intervention. Despite months of rhetoric about Iraq's fledgling nuclear weapons program, the Allied forces, combined with Israel, have always had a monopoly on usable nuclear weapons in the Middle East. In January of 1991, Greenpeace estimated that the U.S. has some 1,000 nuclear warheads in the region: 300 in Turkey and 700 aboard U.S. warships and attack submarines. The Bush administration's unwillingness to rule out the use of nuclear weapons against Iraq still threatens the entire Middle East region, and increases the likelihood of a frightening new nuclear arms race throughout the Third World.

A future attack by Iraq or anyone else on Israel's 20 year old Dimona reactor would be an ecological disaster of global proportions.

The environmental consequences of U.S. intervention will be felt for a long time to come. The military forces in Saudi Arabia have consumed huge quantities of fuel and water, destroyed fragile desert soils with their tank maneuvers, and left waste and sewage equivalent to that of a city of half a million people. The destruction of vegetation will disrupt the water cycle and reduce already meager rainfalls. Joni Seager, in the *Village Voice,* estimated that the troops produced at least ten million gallons of sewage a day. Trash and human wastes are either buried or burned. In addition, a technologically intensive operation like the Gulf War uses and disposes of huge quantities of chemical solvents, paints, acids, lubricants and other toxic materials, in addition to fuel and explosives.

Much has already been written about the military's use of massive quantities of fuel. They use oil for everything from jet fuel to air conditioning, amounting to 80 percent of their total energy use. Even in peacetime (which of course for the military means preparation for war and small "operations" like the invasion of Panama), the U.S. military uses 37 million tons of oil a year, according to the Worldwatch Institute. This is enough to run all the urban mass transit systems in the U.S. for 14 years. In a war, fuel use can multiply five- to ten-fold: during World War II, the Pentagon was responsible for a third of all the energy used in the U.S. economy. Estimates for the Gulf War begin at 20 million gallons a day.

Military vehicles use staggering amounts of fuel. Aircraft carriers average up to 150,000 gallons of fuel a day. B-52's use 3,600 gallons per hour, fighter planes about half that much. According to the Earth Island Institute's Gar Smith, "An F-16 jet on a training mission ignites more fuel in a single hour than the average car owner consumes in two years," well over a thousand gallons. The new M-1 Abrams tanks use 22 gallons of fuel to travel one mile, or almost 300 gallons per hour. Many of these tanks are said to be armed with super-hardened shells made from "depleted" uranium left over from the process of enriching uranium to manufacture nuclear reactor fuel.

WHAT IS EDITORIAL BIAS?

This activity may be used as an individualized study guide for students in libraries and resource centers or as a discussion catalyst in small group and classroom discussions.

The capacity to recognize an author's point of view is an essential reading skill. The skill to read with insight and understanding involves the ability to detect different kinds of opinions or bias. *Sex bias, race bias, ethnocentric bias, political bias and religious bias* are five basic kinds of opinions expressed in editorials and all literature that attempts to persuade. They are briefly defined in the glossary below.

Glossary of Terms for Reading Skills

Sex Bias—the expression of dislike for and/or feeling of superiority over the opposite sex or a particular sexual minority

Race Bias—the expression of dislike for and/or feeling of superiority over a racial group

Ethnocentric Bias—the expression of a belief that one's own group, race, religion, culture or nation is superior. Ethnocentric persons judge others by their own standards and values.

Political Bias—the expression of political opinions and attitudes about domestic or foreign affairs

Religious Bias—the expression of a religious belief or attitude

---

Guidelines

1. From the readings in Chapter Two, locate five sentences that provide examples of editorial opinion or bias.

2. Write down each of the above sentences and determine what kind of bias each sentence represents. Is it *sex bias, race bias, ethnocentric bias, political bias or religious bias?*

3. Read through the following statements and decide which ones represent a form of editorial bias. Evaluate each statement by using the method indicated below:

- **Mark** — (S) for any statements that reflect sex bias.
- **Mark** — (R) for race bias.
- **Mark** — (E) for ethnocentric bias
- **Mark** — (P) for political bias
- **Mark** — (F) for statements that are factual.

___ 1. Ecocide is an apt description of the effects of the U.S. war against Vietnam.

___ 2. Defoliants caused enormous genetic destruction in Vietnam.

___ 3. The Soviet Army had no right to leave behind a mountain of environmental damage in Eastern Europe.

___ 4. The Germans must pay the price for their atrocities against the Russian people.

___ 5. France's sovereignty over French Polynesia, and its nuclear testing program are justified by historic declaration.

___ 6. French and American strategists regard any "nuclear-free zone" in the Pacific as a threat to their national interests.

___ 7. Radioactive contamination remains a serious health threat in French Polynesia.

___ 8. In order to raise quick cash for their war effort, insurgents in Burma are destroying the environment.

___ 9. The military has long considered itself beyond existing environmental laws.

___ 10. The military is in the business of protecting the nation, not the environment.

___ 11. The allies were justified in their destruction of Iraq.

___ 12. Saddam Hussein is to blame for any ecological damage during the Gulf War.

___ 13. Radioactive waste was an unfortunate by-product of the U.S. nuclear program.

___ 14. The U.S. military should be held accountable for all the toxic waste it has produced.

___ 15. Some information on military contamination is too "sensitive" to be released to the public.

CHAPTER 3

WAR AND THE ENVIRONMENT: IDEAS IN CONFLICT

WAR AND THE ENVIRONMENT: IDEAS IN CONFLICT

DESTROYING THE ENVIRONMENT

Center for Defense Information

The Center for Defense Information supports an effective defense and opposes excessive spending for weapons and policies that increase the likelihood of war. The Center publishes a newsletter, The Defense Monitor.

Points to Consider:

1. How has the military actually undermined U.S. national security?

2. Why are the nuclear "bomb factories" so threatening to the environment?

3. Why is cleanup being hampered?

4. Identify the major kinds of military pollution.

Center for Defense Information, "Defending the Environment? The Record of the U.S. Military," The Defense Monitor, Vol. 18, No. 6, 1989.

The pursuit of military power has gravely undermined another element of overall security: the sound, healthy environment necessary to sustain life on this planet.

Americans have only just begun to grasp the full nature and extent of the environmental damage and danger to public health that is the result of more than four decades of military competition between the United States and the Soviet Union. The pursuit of military power has gravely undermined another element of overall security: the sound, healthy environment necessary to sustain life on this planet. In building and maintaining its massive armed forces, the United States has created widespread environmental problems at home and abroad.

The U.S. military establishment collectively is the world's largest industry. It includes 1,246 Department of Defense (DOD) military bases (871 in the U.S. and 375 overseas), 17 major Department of Energy (DOE) nuclear warhead production sites in 12 states, 12 chemical weapons production and storage facilities (2 overseas), and thousands of private companies and university laboratories throughout the United States engaged in military weapons research, development, and production. This military conglomerate is also one of the world's biggest polluters, fouling air, land, and sea with refuse ranging from ordinary garbage dumped from naval vessels to poisonous radioactive and chemical wastes.

No Regulation

From the end of World War II through the runaway military spending of the Reagan presidency, the United States race to possess more and deadlier weapons overwhelmed the concern for a cleaner and safer world. The U.S. military establishment has either ignored or obtained exemptions from laws such as the Resource Conservation and Recovery Act and the Clean Water Act that set environmental and public health and safety standards for private industries, individuals, and municipalities in the United States. The majority of U.S. military facilities do not meet federal state hazardous waste control requirements.

In the 1970s and 1980s, while civilian industries were being forced to adjust to new environmental standards, the military establishment remained focused on its narrow military missions. Its emphasis on isolation, self-regulation, secrecy, and meeting production goals, whatever the costs, contributed in large part to

BULLETINS

50 YEARS TO CLEAN UP TOXIC WASTE DUMPS— GOV'T REPORT

STAYSKAL TAMPA TRIBUNE

I'D QUIT THIS LINE OF WORK IN A MINUTE IF IT DIDN'T HAVE SUCH GREAT JOB SECURITY!"

Reprinted by permission: **Tribune Media Services.**

the current environmental mess. The miliatry has long had its own separate norms of behavior and set of priorities. Respect for the environment has never been one of its virtues. Now, environmental audits, notices of violations, lawsuits, and adverse public opinion may compel the military to shed its bureaucratic myopia, lift its secrecy, and begin respecting the same environmental laws as the rest of society.

Based on current DOD, DOE, and General Accounting Office (GOA) estimates, the total cost of bringing U.S. military facilities into compliance with environmental laws and mending the damage they have caused could easily exceed $150 billion. The real cost of all environmental damage caused by the military establishment, however, cannot be expressed in dollars, because some costs are incalculable and some losses are permanent. The task of decontaminating groundwater alone will be enormous. Congressman Richard Ray (D-Georgia), Chairman of the Environmental Restoration Panel of the House Armed Services Committee, believes "environmental restoration now has to move into the forefront of almost everything else short of a national emergency." It is assured that future generations of Americans will be paying the bills far into the 21st century.

Nuclear Bomb Factories

The most serious military threats to the environment, by far, come from the nuclear energy the United States has used since 1945 to produce nuclear weapons and to propel Navy vessels.

The Department of Energy (DOE) is responsible for nuclear warheads research, development, testing, production, and waste disposal. Recently there have been shocking disclosures of environmental contamination and disregard for public health and safety at 17 nuclear warhead production facilities in 12 states. Hundreds of billions of gallons of extremely toxic radioactive, chemical, and mixed wastes have been discharged into the soil and air in violation of federal hazardous waste disposal laws. Names such as Fernald, Hanford, Rocky Flats, and Savannah River have become synonymous with the military establishment's disregard for the environment.

Hanford

After failing to meet safety requirements, nuclear reactors at the Hanford Reservation in Washington State closed permanently in 1988. Radioactive iodine was deliberately and secretly released from the site in the 1950s. An estimated 200 billion gallons of radioactive and chemical wastes were dumped into ponds and unlined trenches, polluting the Columbia River and area drinking water. Leaking underground tanks containing high-level radioactive waste continue to damage the environment while awaiting cleanup.

Disposal Dilemma

Because plutonium is extremely deadly and remains radioactive for approximately 240,000 years, it must be treated and permanently stored to prevent its dispersal. Disposal of radioactive waste from nuclear weapons production is perhaps the most serious long-term environmental challenge resulting from military programs. The DOE annually produces hundreds of millions of gallons of chemical, radioactive, and mixed waste.

For more than four decades radioactive waste products have been shunted to various "temporary" storage sites while scientists and engineers continue to look for a safe means of permanent disposal.

POLLUTION ON COMMAND

Fairchild Air Force Base, west of Spokane, Washington, has been used since 1950 to maintain and repair aircraft, primarily bombers and tankers. The site holds more than 4,000 drum equivalents of carbon tetrachloride and other solvents, paint wastes, and plating sludge containing heavy metals. Approximately 400 private wells serving about 20,000 people are within three miles of the facility.

Carl Lind, "Pollution on Command," **Not Man Apart**, June/July 1990

Other Military Threats to the Environment Include:

- Military space programs are also a source of radioactive pollution. Over the past 30 years the United States and the Soviet Union have launched more than 60 spacecraft with nuclear power sources aboard, approximately 15 percent of which suffered some form of failure or accident.

- In 1989, the Senate expressed concern that the Department of Defense (DOD) has devoted too little money and effort to finding ways to comply with nuclear arms reductions in "an environmentally benign manner".

- There are approximately 150 public and private U.S. chemical and biological warfare (CBW) laboratories in more than 30 states and 15 foreign countries that are of questionable safety and may be vulnerable to theft or sabotage.

- In February 1988, Maryland environmental officials reported finding 89 hazardous waste violations within a 15-month period at the Aberdeen Proving Ground, an Army facility near Edgewood, Maryland, where chemical weapons have been developed and tested for decades.

- For over 40 years, the Rocky Mountain Arsenal near Denver, Colorado, was a chemical weapons production facility. According to the Army, surface soil and water at 80 different sites covering as much as a quarter of the complex have been contaminated.

- In 1988 Army officials told a congressional subcommittee that the United States' old chemical weapons stockpile (produced before 1969) includes "more than 1,000" leaking weapons and that the problem is likely to worsen in future years.

- Miiltary biological defense research continues to pose the danger of a serious accident or epidemic. The Army's Biological Defense Research Program (BDRP) involves deadly bacteria, viruses, toxins, and recombinant DNA technology. Whereas a bullet eventually comes to a rest after it is fired, a microbe can multiply and spread infection to humans perhaps indefinitely. A virulent disease could be carried unwittingly out of a test facility and spread to family, neighbors, community, and then possibly beyond state and national borders without anyone knowing the origin or nature of the disease.

- The Department of Defense estimates that the military services produce about 400,000 tons of hazardous waste each year. Some place the figure as high as 500,000 tons—more than the top five civilian chemical companies combined.

- Perhaps the most obvious environmental cost of the military is its inordinate consumption of non-renewable natural resources. The military uses vast quantities of non-fuel minerals such as copper, lead, tin, titanium, cobalt, germanium, vanadium, chromium, and manganese. It has been estimated that the fuel consumed by the DOD in a single year would run the entire U.S. public transit system for 22 years.

Summary

- Production of nuclear and chemical weapons during the past 45 years has caused enormous environmental problems in the United States.

- Hazardous radioactive and chemical wastes threaten the health of the American public.

- Continued production of nuclear and chemical weapons is further aggravating damage to the environment.

- There are more than 15,000 suspected contaminated military waste dumps spread across the United States.

- The Department of Energy's nuclear bomb factories have created millions of tons of highly radioactive wastes.

- No proven method has been devised for the safe and permanent disposal of nuclear wastes.

10 WAR AND THE ENVIRONMENT: IDEAS IN CONFLICT

PRESERVING THE ENVIRONMENT

William H. Parker

William H. Parker made the following statement in his capacity as Assistant Secretary of Defense for the Environment.

Points to Consider:

1. How many military installations exist within the United States? How many exist worldwide?

2. What are the major environmental concerns of the Defense Department?

3. How are government owned, contractor operated (GOCO) facilities defined?

4. Why is private ownership related to the environmental concerns of the Defense Department?

Excerpts from testimony by William H. Parker before the House Committee on Armed Services, May 17, 1989.

The bottom line is that if we don't deal with these environmental issues properly and up front, the national security can be eroded.

I welcome the opportunity to acquaint you with the ways in which the Department is committed to environmental enhancement and, in particular, to expand on our efforts through the Defense Environmental Restoration Program for the cleanup of past hazardous waste sites.

At the DOD level, we're involved with five things. First, the establishment of centralized policy. Second, where appropriate, providing consistency in our programs. Third, problem solving throughout DOD and outside of DOD. Fourth, providing strategic planning for the future in our environmental programs. Fifth, is the Defense Environmental Restoration Program for managing our hazardous waste programs.

Next, you'll see our major components: Army, Navy, Air Force, Defense Logistics Agency, as well as other agencies within DOD. Below that are decentralized operations or installations. These are the superstars. We're proud of them. We look on them as our heroes. They're the ones that have the responsibility for compliance in the field. These people are literally the ones who are where the rubber hits the road.

Also I'd like to discuss with you centralized DOD policy. Here are four matters that I'd like to call to your attention. First of all, we are responsible for compliance with environmental laws. Second, we maintain environmental stewardship for land, air, water and natural resources, all of which I'll be discussing today. The third is cooperation, not only cooperation within the Federal agencies, but at the State and local government agencies and outside of the United States with other national governments. Fourth, policy analysis, and that's analysis of existing laws as well as proposed laws.

Now I'd like to turn to DOD operations. We are made up of mega-industries. We also are communities who have complex challenges. We are not a private industry nor are we municipalities. We are governed, managed and financed differently from the private sector. We also have a different primary mission.

Military Installations

This map shows our major military installations within the United States. We have approximately 900 of those installations

Recycling operation aboard the U.S.S. Lexington. Defense Dept. photo.

and I believe that the State of Vermont is the only state where a major installation is not located. Worldwide, by the way, we have about 1,100 installations. Within the United States, we have to deal with both multi-federal agencies as well as multi-state agencies. Every state and every installation has its own set of challenges, rules, regulations and laws.

I want to brag a little bit about our proactive leadership role in cleaning up hazardous waste sites. Today, certainly hazardous waste is our biggest problem. The Resource, Conservation and Recovery Act (RCRA), for example, relates to the cleanup of our current operations. It's management intensive, and it has very complicated rules.

One of the keys is waste minimization and minimization of the use of hazardous materials. We feel this is key to our future. We want to keep these materials out. We do not want to get in the position where we are creating problems for tomorrow. We have enough of yesterday's problems to handle.

Part of what we're doing is reuse and recycling. We started this in 1980 which is four years before any law was enacted that required the private sector to look at reuse and recycling. We continue to be a leader in this area.

Perceptions

One thing that concerns us daily is the perception that DOD could be uncooperative. We can't live with this. We have to be looked on as someone who's trying to solve the problem. We do intend to do the right thing, and we do obey the law. But I do want to again emphasize, it's going to take a long time to fix some of these problems that were created as far back as before World War II.

The bottom line is that if we don't deal with these environmental issues properly and up front, the national security can be eroded. Our strategies in dealing with this are to better communicate both internally and externally, be smarter managers, integrate environmental requirements into all of our activities and provide flexibility in the field.

Congress has been of great help in the past. For that I want to thank you. We look forward to working with you. I think that you can see there are a number of challenges. We appreciate your interest. There is one thing that I do want to leave you with when we talk about these 900 installations. This is where our people live, and where they work. We are very concerned about their well-being.

GOCO Facilities

Government owned, contractor operated (GOCO) facilities are an extremely important component of our overall national security infrastructure. We currently own 66 plants (Army-34, Navy-18, Air Force-13, Defense Logistics Agency-1) located in 31 states. They are as diverse in their function and type of contract as their geography. Although these facilities represent a major investment, we do not generally believe that government ownership of industrial property is appropriate. There are exceptions, such as GOCOs involved in production of military-unique products like ammunition. The department therefore has a major initiative to transfer ownership of GOCOs to the private sector. We have divested over 21 facilities in the last ten years and have plans to transfer as many additional facilities as possible to private ownership. When we relinquish ownership we must meet all environmental requirements.

Today's environmental regulations make property transfers an extremely complex business and sales more difficult. We are not attempting to evade environmental cleanup responsibilities by selling these facilities. As former owners, we still retain our liability as a potentially responsible party. Our experience has been that plants transferred to the private sector will be modernized by the new owners in order to become more competitive in the market place. These modernization efforts should tend to reduce the environmental effects of the plants' operations, avoid DOD costs associated with ownership and return the properties to local tax bases.

Another important benefit of private ownership is that it affixes responsibility for environmental compliance with a single party and fosters compliance. The plants that become contractor owned, contractor operated (COCO) facilities will be under complete control of the contractor. For the most part, contractors will not be able to claim that the government is limiting their ability to operate their facility in compliance with environmental requirements.

Environmental Compliance Responsibility

Our policy is that all DOD facilities and their operators, including GOCO plants, must comply with environmental requirements. As a demonstration of our commitment, the DOD has invested approximately $112.6 million (Air Force = $63.8 million; Army = $44 million; Navy = $4.8 million) of non-defense environmental restoration account (DERA) funds over the last three fiscal years on projects to obtain and maintain environmental compliance at GOCOs.

Our contracts with GOCOs also include a standard federal acquisition regulation (FAR) clause which specifically refers to the Clean Air Act and Clean Water Act. We believe this is sufficient to sensitize the contractor to perform consistently with other environmental laws. Regardless of whether or not contractual requirements exist, environmental compliance responsibilities are imposed directly on operators by existing federal and state statutory and regulatory requirements.

Challenges

The concepts of liability and responsibility reflected in recent environmental law create many complicated problems. For one example, it has been suggested that environmental problems resulting from activities conducted during World War II, under the War Powers Act, are the sole responsibility of the federal government. When the scope of the massive industrial activity which was conducted during the War is realized, it is difficult to believe that drafters of environmental legislation which established a "once liable always liable" principle, intended for the government to assume responsibility for all the problems which may have been created by those activities. The specter of litigating these claims through the courts in an effort to clearly affix 40 and 50 year old liability is not a pleasant national prospect.

The department funds most cleanup costs for sites created by past practices. In effect, the responsible contracting officers in the DOD components have determined that the operating contractors were using facilities provided by the government in ways commonly accepted as reasonable business practices at the time the contamination occurred. Most of these funds were used to find and fix contamination at these facilities. If we subsequently determine, however, that the contamination was caused by illegal activities or grossly negligent contractor operations, we will seek reimbursement.

Summary

The department takes its environmental compliance responsibilities seriously. We are making significant investments to meet our commitments in our capacity as "landlord." We believe our contract operator "tenants" also have compliance responsibilities. Given the complexity of current environmental law and regulations and the differing contractual relationships at each of our GOCOs, sorting out the degree of financial responsibility in a particular situation is difficult and must be made on a case by case basis.

The complexities of the "landlord-tenant" relationship at our GOCOs also present difficult issues for environmental regulatory enforcement agencies. The EPA has recognized these complexities and is working to develop an enforcement strategy.

11 WAR AND THE ENVIRONMENT: IDEAS IN CONFLICT

WAR AND ECOCIDE: THE POINT

Matthew Nimetz and Gidon M. Caine

Matthew Nimetz and Gidon M. Caine are lawyers practicing in New York City; Nimetz previously served as Under Secretary of State.

Points to Consider:

1. How did Saddam Hussein violate international law?

2. What existing laws govern ecological destruction during war?

3. How should Saddam and Iraq be punished?

4. How can the rules of war be used and improved to protect the environment?

Matthew Nimetz and Gidon M. Caine, "Crimes Against Nature," **Amicus Journal,** Summer 1991.

Only during the past fifty years have we mastered technologies that permit us to cause widespread and long-term damage to our earth.

When the United Nations Security Council condemned Iraq's invasion of Kuwait, it also condemned Saddam Hussein's deliberate assault on the environment. But should it go further? How can the community of world governments punish Hussein for threatening the biosphere, and not just Kuwait's piece of it? More importantly, how can the world community prevent future Saddam Husseins from threatening the regional and global environment yet again?

International law already prohibits the destruction of the environments of the nations at war during conflict, but there is no law prohibiting harm that crosses national boundaries. To address this broader problem of damage to the global commons, the community of nations must develop a consensus on what cannot be done to the biosphere during wartime; the international community should adopt an agreement designed to reduce war's environmental harm, and should create an organization similar to the Red Cross to report on the warring parties' adherence to those norms.

Iraqi Liability

The Persian Gulf conflict provides an opportunity to engage in a serious review of the laws governing the environmental toll of warfare. Just days after the war began, Iraqi forces started channeling Kuwaiti oil from the Sea Island Terminal into the Gulf, causing a spill that amounted to six to eight million barrels of oil—the largest oil spill in history. In early June, 42,000 gallons of oil were still flowing into the Gulf each day, and a 300-mile strip of northern Saudi Arabian coastline had been covered with oil.

In the desert, the amount of oil being released into the air and onto the land was considerably greater. At least three million barrels of oil went up in smoke every day over Kuwait from oil well fires ignited by Iraq in February, in what a *New York Times* editorial called "an act of insane vindictiveness". In four days of bombing, Iraq blew up most of Kuwait's 1,250 oil wells, setting almost 600 wells afire and flooding large areas of the country with oil.

The United Nations Security Council, by a vote of twelve to one (with three abstentions), adopted Resolution 687 on April 3,

1991, which reaffirmed Iraqi liability for "environmental damage and the depletion of natural resources. . .to foreign governments, nationals, or corporations," resulting from Iraq's invasion of Kuwait. The resolution also created a fund to collect approximately a quarter of Iraq's annual oil revenues to pay claims expected to reach $50 billion.

The Rules of War

While Resolution 687 is a good first step, it fails to address whether Iraq should pay for the long-term effects of the ecological disaster in the Gulf on the biosphere; the resolution focuses instead on the identifiable damage to the environment of each country in the region. Compensation should cover the short-and long-term harm done by Saddam Hussein's military strategy to countries far from the Gulf, and such devastation must be prevented from happening again.

History teaches that all warfare affects the environment. Carthage had its fields plowed with salt by Roman armies; the Dutch saw dikes breached; and the South felt the burn of Sherman's march through Georgia. But only during the past fifty years have we mastered technologies that permit us to cause widespread and long-term damage to our earth.

One of the more important statements on environmental damage during wartime is the Convention on the Prohibition of Military or Any Other Hostile Use of Environmental Modification Techniques (known as the Enmod Convention), signed on May 18, 1977. This convention bans environmental modification techniques that have "widespread, long-lasting, or severe effects as the means of destruction or injury to any other. . .Party to the Convention."

The issue of environmental warfare was discussed in the 1977 Protocol I Addition to the 1949 Geneva Convention, the major international agreement on accepted practices of war. The 1977 protocol prohibits states from employing "methods or means of warfare which are intended, or may be expected, to cause widespread, long-term and severe damage to the natural environment."

Whatever the limitations of the Enmod Convention and the 1977 Geneva Protocol I, they have real meaning and are sufficient to impose liability on Iraq for its actions in Kuwait and the Gulf. However, using the Enmod Convention simply to compensate Kuwait and to clean up the Gulf does not address the fundamental wrong done by Iraq to the global environment. This harm cannot be adequately described as damage to any one nation or people.

The law of war is still at a rudimentary stage, particularly in dealing with environmental impacts. We should use the Gulf War as an occasion to develop a mechanism to compensate the earth itself for harm done during warfare; the mechanism must recognize that the preservation of the biosphere during wartime can no longer be simply the responsibility of the belligerents involved.

Punishable Acts

The world community must articulate the growing consensus that those who wantonly and significantly damage the biosphere during war in violation of international law are outlaws, not merely technical violators of agreements. In flagrant cases, the personal culpability of the individuals responsible should be considered. If the world community deems Saddam Hussein's actions to be truly outrageous, these acts should be punishable not merely as violations of treaties, but as war crimes. If the world community cannot reach a consensus on that issue, at the very least it should do more to protect the biosphere from environmental damage during wartime.

Going beyond the Gulf War, we make three recommendations. First, the parties to the convention should convene to discuss the environmental effects of the Gulf War and ways to strengthen the convention itself. Second, we need an equivalent of the Red Cross, perhaps ultimately a corps of scientists and experts wearing green arm bands, to inspect and report during conflicts on behalf of nature itself. Third, every nation should sign and ratify the Enmod Convention immediately.

Finally, it is time for the United States and other western countries to ratify the Geneva Protocol. The 1977 Geneva Protocol I contains a number of important provisions governing modern warfare. During the Vietnam era, the United States was reluctant to adopt this convention for fear that it would be applied to its use of defoliants.

The global environment is at risk because the technology of war allows us to put it at risk. The conflict in the Persian Gulf has reinforced that lesson. Global diplomacy, already a force in addressing global warming and ozone depletion, provides us with a way of reducing, if not eliminating that danger—but the time to act is now, not after some other leader with an army at his command engages in gratuitous environmental vandalism somewhere else.

Our children should not condemn us for having seen the danger and having done nothing. That, and a crippled environment, should not be our legacy.

12 WAR AND THE ENVIRONMENT: IDEAS IN CONFLICT

WAR AND ECOCIDE: THE COUNTERPOINT

Zoltan Grossman

Zoltan Grossman is a Wisconsin-based researcher of military interventions and is the outreach director of the National Committee Against Registration and the Draft.

Points to Consider:

1. Who is to blame for ecocide in the Gulf War?

2. What serious eco-threats resulted from U.S. bombing?

3. How does the U.S. apply a double standard with regard to weapons of mass destruction?

4. What solution is suggested by the author?

Zoltan Grossman, "Ecocide in the Gulf," Z **Magazine**, March 1991.

While much attention has been paid to Iraq's destruction of Kuwaiti oil facilities, less has been paid to the U.S. bombing of Iraqi oil refineries, rigs, tankers, and other targets, resulting in widespread spills.

The world has been horrified by the ecological devastation resulting from the Gulf War. On the weekend of January 24-27, Iraq's release of oil from a Kuwaiti facility caused a massive flaming oil slick along the Gulf coast, and threatened the Saudi water supply. Saddam Hussein's alleged "scorched water strategy" was called "kinda sick" by President Bush, and the U.S. Air Force proclaimed itself the guardian of the environment by bombing the facility's valves.

But wait a minute. U.S. condemnation of Iraq's oil spill somehow implies that U.S. forces have not been deliberately targeting the environment. In fact, both sides have seen the destruction of the environment—and thus of the health of the civilian population—as a mission objective in the Gulf War.

Bush's claim that U.S. surgical strikes are not targeting civilians is directly contradicted by his choice of sites to destroy. Civilians don't have to be directly hit by air strikes in order for many civilians to die. The targets have included water purification plants, water desalinization plants, oil rigs and refineries, nuclear reactors and laboratories, chemical plants, and biological facilities. It is virtually impossible for civilians to have not been contaminated with germs, chemicals, or radiation in the air or water as a result of these bombings.

Why don't we know more about the effect of the bombing campaign? The interests of the U.S. and Iraqi leadership coincide in covering up the number of civilian casualties. For the U.S. military, a high civilian death toll would be bad for public relations; for the Iraqi military, it would be bad for public morale.

Disastrous Results

To see the disastrous impact of bombing of the civilian population, one only has to look at a U.S. Air Force Tactical Pilotage Chart of Central Iraq. Thin strips of green follow the valleys of the Tigris and Euphrates Rivers, converging in a larger green patch in Iraq's heartland around Baghdad. This densely populated fertile region is surrounded by wetlands, corresponding to the ancient extent of the Persian Gulf. The wetlands, in turn, are surrounded on all sides by desert. The region's biosphere is reliant on a very thin lifeline, and it is along

this lifeline that the most contaminants have been released by the bombings.

While much attention has been paid to Iraq's destruction of Kuwaiti oil facilities, less has been paid to the U.S. bombing of Iraqi oil refineries, rigs, tankers, and other targets, resulting in widespread spills. At the same time, bombers have knocked out the civilian water supply to major cities like Baghdad (with a population of three million) and Basra, and bombed water purification plants. Civilians are reliant on treated water from the two major waterways. Foreign worker refugees have reported that Baghdad civilians are now drawing water directly from the Tigris, and that signs of water-borne diseases (such as acute diarrhea) are appearing in children.

Weapons of Mass Destruction

It is not known what kind of contaminants might have leaked into the water supply from the destruction of chemical and biological facilities. Iraq has used chemical weapons on two fronts—mustard gas against Iranian troops, and nerve gas against Kurdish civilians. At least 11 Iraqi sites have been bombed because of the possibility they may be producing bio-chemical warfare agents. But is the cure worse than the disease? When a small section of Union Carbide's Bhopal fertilizer plant leaked in 1984, more than 2,000 Indians died from inhaling methyl isocyanate gas. What is happening to the surrounding civilian population where Iraqi bio-chemical plants are being deliberately set ablaze by U.S. bombs?

Among the targeted facilities have been a fertilizer plant in Basra (Iraq's second largest city, with a population of 371,000) and the Quaim phosphate mine near Syria (organophosphates are a key ingredient in both fertilizers and nerve gas). Also on the bombing list have been an ethylene plant in Musayyab (pop. 16,000), and a facility in Fallujah (pop. 36,000), both along the Euphrates River. Along the Tigris River, bombers hit a plant that may have made nerve gas precursors in Sammarra (pop. 25,000), the Salman Pak facility which allegedly made biological warfare agents, and a facility in Bayji (pop. 7,000). One Cable News Network (CNN) reporter described one of the burning chemical plants as sending off "green flames". All told, at least 600,000 Iraqis live in towns (not including Baghdad) that may have been covered by toxic gases or germ clouds from the bombings, where no foreign reporters were present.

A similar threat faced civilians living near Iraqi nuclear facilities destroyed by B-52 strikes—the first successful strikes against "hot" nuclear facilities ever in warfare. Before the war, Iraq was allegedly engaged in building a nuclear device with success anywhere from one to ten years away, depending on who you talked to. When the Israeli Air Force bombed the Osirak atomic reactor within the Tuwaitha nuclear complex near Baghdad a decade ago, radioactive materials were not present within the facility. Yet before the U.S. destruction of two Tuwaitha reactors, the facility used enriched uranium, which had been recently inspected by the UN's International Atomic Energy Agency (IAEA).

Double Standard

A certain hypocrisy permeates the current debate around weapons of mass destruction. The United States is the only country to have used nuclear weapons in warfare, and has threatened their use over a dozen times since. The U.S. also used chemical weapons by spreading Agent Orange over Indochinese forests and croplands, in the process poisoning untold numbers of civilians and U.S. soldiers. Yet President Bush is appointing himself as the guardian against Iraqi chemical arms and potential nuclear arms.

Bush does not apply the same criteria to Israel, which the CIA confirms as possessing an arsenal of atomic bombs, and whose officials recently admitted possessing chemical arms. Though Iraq's capability has been destroyed, it is only a matter of time before another Islamic nation tries to counter Israeli weapons.

Instead of destroying the weaponry of one side in the Arab-Israeli dispute, causing untold destruction, doesn't it make sense to initiate regional disarmament? All facilities in the Middle East making weapons of mass destruction could be dismantled under UN supervision, and the region enshrined in treaties as a zone free of nuclear, radioactive, and biological arms of any country. By proposing such an agreement, the U.S. could have disarmed Saddam Hussein without resorting to an air assault, but it chose not to in order to preserve Israel's strategic advantage. Such treaties could provide a precedent for other world regions where border disputes could all too easily slip into technological genocide.

EXAMINING COUNTERPOINTS

This activity may be used as an individualized study guide for students in libraries and resource centers or as a discussion catalyst in small group and classroom discussions.

The Point

Much of the environmental damage resulting from the Persian Gulf War was an unfortunate side effect of modern warfare. Similar damage has resulted from every major war fought in our time and is one price we must be willing to pay for the greater good of national security and world peace. While this certainly does not justify any deliberate acts of eco-terrorism, we must be ready to accept a certain amount of environmental damage that is inevitable during full-scale air, land and sea operations. It is unreasonable to apply commercial "air quality standards" or other environmental regulations to the military in times of war.

The Counterpoint

The entire concept of modern warfare as a means to solve international disputes is ridiculous when we realize the suicidal effect war has on our global environment. Peace and national security are threatened by the terrible legacy of the arms race, warfare, and the resultant destruction of our planet. Only since World War II have we mastered the technology to literally destroy ourselves in modern war. The death, destruction and ecological devastation unleashed during the Gulf War, repeated and multiplied in the name of "security", will surely lead to global insecurity. It is unreasonable to consider such military operations as solutions to anything.

Guidelines

Social issues are usually complex, but often problems become oversimplified in political debates and discussion. Usually a polarized version of social conflict does not adequately represent the diversity of views that surround social conflicts.

1. Examine the counterpoints above. Then write down possible interpretations of this issue other than the two arguments stated in the counterpoints above.

2. Which argument do you agree with the most? Why?

13 WAR AND THE ENVIRONMENT: IDEAS IN CONFLICT

MILITARY DESTRUCTION OF PUBLIC LANDS

Michael A. Francis

This reading examines the military use of public lands in the U.S. and the threat to environmental and human resources. Michael A. Francis is counsel on National Forest Issues for the Wilderness Society.

Points to Consider:

1. What is the most troubling aspect of the military impact on public lands? Why?

2. How is wildlife affected by military exercises? Discuss several examples.

3. In what ways does the military threaten the National Parks and National Forests? Be specific.

4. Why are overflights so dangerous?

Excerpted from testimony submitted by Michael A. Francis before the House Subcommittee on National Parks and Public Lands of the Committee on Interior and Insular Affairs, January 3, 1990.

In Nevada, the Navy admitted that over a period of years, its aircraft had accidentally dropped thousands of bombs outside the 22,000-acre bombing range near Fallon Naval Air Station.

Military activities on public lands are not new. The military services have had access for many years to both public land and airspace throughout the United States, particularly lands managed by the Bureau of Land Management (BLM), the U.S. Forest Service (USFS) and the Federal Wildlife Service (FWS) in the West.

Recently, however, the military has proposed a series of extraordinary expansions of its training domain that combined would have an enormous impact on the many important values public lands provide. These expansions would add millions of acres to the more than 25 million acres already controlled by the Department of Defense. The boldest of these plans, involving Montana, Idaho, Utah, and Nevada, would quadruple the already sizeable airspace that the military has restricted to its own exclusive uses in those states. How much is enough?

The military argues its acquisitiveness is necessary because of: 1) technological advances in weaponry that require more training over larger areas and, in the case of aircraft, at lower altitudes; 2) the closure of certain U.S. bases around the country and 3) the growing opposition of Europeans to our overseas exercises.

Few would dispute the importance of providing sufficient space for our troops to prepare for combat. Moreover, we recognize that the military cannot conduct exercises in heavily populated areas. The sad fact is that there is less and less open space in this country, and the competition for use of it grows fiercer every day.

Even so, controls on the military's appetite for the use of this public land must be set. How free are we if we sign over more and more of our natural treasures for use by tanks, helicopters, and bombers? The Pentagon likes to describe sonic booms and other military aircraft noise as "the sound of freedom". Beyond a certain point, however, it becomes the sound of freedom lost.

Military Impact

The use of land and airspace by the military damages public lands and the wildlife that depends upon them for habitat. The impact includes:

Public lands in Great Basin National Park, Nevada.
Photo by Ron Swanson

- Erosion and other damage to sensitive soils and watersheds.
- Disturbance and death of wildlife.
- Disruption of human activities, including risk to human safety.

Erosion and Other Damage

According to a recent study, army units from Fort Carson did considerable damage to southeastern Colorado grasslands used in 1986 and 1987. Live plants had covered 22 percent of the area before training began, but covered only nine percent two years later. While this acreage was not in the public domain, this experience is a sign of what can be expected.

Disturbance and Death of Wildlife

In 1988, the Bureau of Land Management (BLM) allowed National Guard maneuvers on some of its Utah lands. According to the BLM, the Guard proceeded to violate 35 of the 81 stipulations on allowable use. A BLM spokesman in the Salt Lake district told *The Los Angeles Times*, "We saw some problems as they were occurring, and we tried to do something to stop them, but the exercise was so large and the authority so scattered that we couldn't find anyone who would be accountable."

One example of the impact on wildlife was destruction of a winter hunting ground for bald eagles by a fire ignited by artillery shells.

Possibly the most troubling aspect of military activities on public lands and within public airspace is the general lack of knowledge or ongoing research about the impacts from those activities. Given the fact that some military operations and overflights have been underway for nearly half-a-century and given the clear requirements for evaluation of environmental impacts from new, expanded, long-term and many other classes of federal activities—it is disturbing to us that little is known and almost no long-term, comprehensive studies have been initiated by the agencies in the Defense Department.

Disruption of human activities

Two of the most significant features of our public lands, especially in the wide open West, is the opportunity for quiet and solitude for outdoor recreation. Increasingly, Americans need to escape the fast pace and the noise of urban areas.

SIERRA CLUB

Below are several examples of what is documented, according to our survey, as having occurred when military activities were conducted on public lands within the United States:

* The Army was asked to drop its use of the National Forests in Florida in 1984 after abuse of those lands which included non-retrieval of human wastes and field ration cans; littering; unattended campfires which became wildfires; cutting of live timber; and pointing and simulating firing of weapons at occupants within the forest.

* Hikers and forest users have been "captured" as "enemies" in Florida and North Carolina by maneuvering troops. Private properties unrelated to maneuvers have been stormed.

* In Georgia, forest monitors within conservation groups have found litter, shell casings, flare canisters, and field ration wrappings. Fox holes and bunkers had been constructed, as well.

* In Mississippi, designation of Black Creek as a wild and scenic river was threatened by machine gun and tank fire all through the night due to National Guard training in the DeSoto National Forest.

* The Konza Prairie Nature Preserve in Kansas, a rare example of one of the few remaining tallgrass prairies in North America which draws visitors and students from all parts of the country, sits approximately seven miles from the Fort Riley training base at Manhattan, Kansas. On maneuver days, the shelling is louder than thunder and can be heard on any part of the Konza Prairie.

* In Nevada, a rancher has reported that low-level helicopter flights drove 40 head of livestock from their water sources, ultimately resulting in their deaths.

* No Mans Island, three miles off the coast of Martha's Vineyard, Massachusetts, is a wildlife sanctuary used for bombing practice by the Navy. Pieces of dead birds can be found scattered throughout the island.

* The Navy plans to conduct low level training flights off Cape Lookout National Seashore, North Carolina, a major environmentally sensitive recreational and wildlife area.

Excerpted from testimony submitted by the Arkansas Chapter of the Sierra Club before the House Committee on Interior and Insular Affairs, Jan. 3, 1990.

However, visitors to designated wilderness areas in Washington State's North Cascades too often see and hear military aircraft.

This is not an exclusively western problem. Nearer at hand is the example of Prince William Forest Park, just south of Washington, D.C. During World War II, 4,862 acres of the park were placed under permit to the Department of the Navy for use as part of the Quantico Marine Base. The permit was renewed in 1958, and in 1972 a permit was issued granting the Navy permission to continue to occupy and use 4,514 acres, with only 348 acres to be returned to the National Park Service.

In 1979, the Director of the National Park Service wrote to the Commandant of the Marine Corps asking to discuss return of the 4,514 acres. However, in 1984 a new special use permit was signed, requiring the Secretary of the Navy's consent before use of the land could be returned to the Park Service. In other words, 40 years after the original need for the Marines to use this land ended, one-quarter of the acreage within the boundaries of Prince William Forest Park is now permanently part of the Quantico Marine Base. Fewer acres mean less opportunity for urban residents to enjoy this outdoor setting.

Just as important is the issue of public safety. In the California desert in 1989, a pair of Navy jets accidentally dropped thousands of bombs outside the 22,000-acre bombing range near Fallon Naval Air Station. A sweep collected about 2,000 bombs, some of them still live. The public became aware of the situation only because a television station learned of it and aired the story.

At a time when East/West tensions seem to be easing, it is ironic that so many of the most special parts of our country are now being targeted for greater military activity. The Department of Defense says that the West Germans are no longer willing to put up with life on a practice range. We don't believe that Montanans or other Americans are, either. Understandably, rural citizens are not happy that they are being asked to accept the hazards of military maneuvers, toxic waste dumps, and other dangerous activity that no one else wants.

14 WAR AND THE ENVIRONMENT: IDEAS IN CONFLICT

THE MILITARY SAFEGUARDS PUBLIC LANDS

Robert A. Stone

Robert A. Stone is Deputy Assistant Secretary for Installations with the U.S. Department of Defense.

Points to Consider:

1. How does the Defense Department safeguard the environment?

2. What cooperative steps are taken by the military in this effort?

3. Why does the military need more land for training?

4. How will base closures increase the need for land withdrawal?

Excerpted from testimony by Robert A. Stone before the House Subcommittee on National and Public Lands of the Committee on Interior and Insular Affairs, January 3, 1990.

The Department of Defense is, and will continue to be, a good custodian of the environment.

To summarize my statement today, I would just like to make three fairly simple points. The first point is that we are custodians of a lot of land, 25 million acres. I did some research to try and figure out what 25 million acres amounted to because it is a number not many of us can comprehend.

I certainly couldn't. It is twice the size of San Bernardino County, California. It is half the size of the state of Minnesota, and it amounts to three-and-a-half percent of all federal lands owned by the United States. We know that it is a precious and irreplaceable national resource.

We work hard to minimize adverse impacts on the land that we control and on the nearby communities and to foster shared use. We preserve endangered species on our land and all sorts of wildlife thrive on it. We are proud of the job we do, preserving the public land.

The second point I would like to make is even more important than our custody of these 25 million acres. It is our custody of four million young Americans. Our primary job is to make sure that none of them is ever killed in combat because we failed to train him or her properly or failed to train their leaders. This job requires lots of land and lots of air space. It requires ground and air forces exercising against other forces posing as the enemy.

The third point I would like to make is that our job is everchanging, as weapons change, as the world political situation changes, and the nature of warfare changes. We are working with the Bureau of Land Management on additional withdrawals that are important to us now, and we will work with them and with the Forest Service and with the Congress in the future as we expand to meet whatever our future needs are.

We know the change is coming. Examples of this change are cruise missiles that didn't exist 10 years ago, that we test today requiring vast amounts of land. Aerial maneuvers used to cover 5 miles. Now they cover 40 miles. In the future they will cover 80 miles. Army units, like mechanized infantry battalions, used to need 4,000 acres to train. Now they need 80,000.

Concerned Custodians

As the custodian of military installations covering 25 million acres, the Department of Defense is deeply concerned with the

The Air Force on patrol over Alaska. Defense Dept. photo.

proper management and environmental protection of public lands. These lands are valuable resources for the present and future national security. We use these lands primarily to conduct training and testing of units, tactics, and weapons systems. The Department uses its land and airspace to train our soldiers, sailors, marines, and airmen to survive in combat.

Realistic training requires large areas of land and airspace so combined ground and air forces can fight against other forces posing as the enemy. To provide this realism in training and testing, we need the ability to change, for the nature of warfare is constantly changing. Adequate weapons testing and training requires sparsely populated areas to minimize the risks to human health and safety.

The Department of Defense is, and will continue to be, a good custodian of the environment. Our management of land has secured and even enhanced habitat for endangered species and other wildlife. We believe that proper environmental planning and protection are of paramount importance to the care of our public lands.

To implement this policy we have negotiated several memoranda of understanding with agencies like the Fish and

Wildlife Service for the North American Waterfowl Management Plan. We have cooperative agreements with host states to manage wildlife at military installations all over the country. We have made arrangements with the Federal Aviation Administration and the Forest Service to improve environmental planning when we need to use airspace and national forests.

We also work with the Department of Agriculture to determine appropriate research for pest and natural resource management. DOD even obtained congressional approval to start environmental volunteer and cost/sharing partnership programs on military installations.

One symbol of our commitment to taking care of our lands is the annual Secretary of Defense's Natural Resource Conservation Award. This is the oldest conservation award in the federal government. Since 1962, the Secretary of Defense has presented this award to the installation that best exemplifies a true commitment to safeguarding our natural resources. This is a highly regarded award and one which commanders zealously seek for their installations.

Training Our Troops

The military services have increasing concerns about their ability to train and test properly. They've begun work on requests for additional withdrawals which are very important to the Department of Defense. These withdrawals represent our current needs for expansion to meet today's training and testing requirements.

Authorizing withdrawal of public lands for military purposes is a difficult task. We think that the Bureau of Land Management (BLM) does a good job balancing local, national, public, and private interests. For withdrawal requests of over 5,000 acres, the Congress gets involved. This system of checks and balances serves the nation well and should be continued.

The Defense Department agrees that lands having multiple use features should be managed by the BLM. Lands designated for military use should, however, be managed by the military. That management includes offering land for multiple use when it is compatible with military operations, planning for environmental concerns, safeguarding endangered species, and using airspace properly. Our policy is, and will continue to be, that of being a good neighbor.

Meeting Future Needs

Of equal, or perhaps greater, importance is retaining the flexibility to expand in the future. Technology has surpassed our expectations just in the last decade. We don't know what the future will bring. The only thing we do know is that there will be change.

The Defense Secretary's Commission of Base Realignment and Closure recognized the need for expanding our training and testing areas, even though their charge was to close bases. The Commission also cited the negative impact that encroachment has on existing installations.

For example, there is competition between developers and the military over deep-water ocean ports. This valuable land is needed by the military to stage personnel and equipment for mobilization. The Commission recognized that the National Training Center at Fort Irwin, California, could not properly train battalion, brigade, and division-level forces with their artillery, missile, and air support. The Commission recommended creation of a larger joint-training area to optimize military use of restricted land, air, and water space.

Securing the defense of our nation includes being responsible to those who serve. The expansions we have planned today are needed to ensure that our young men and women are as well trained as possible. But we also must retain the flexibility to expand in the future.

15 WAR AND THE ENVIRONMENT: IDEAS IN CONFLICT

NUCLEAR WEAPONS PLANTS ARE DANGEROUS

H. Jack Geiger and Tony Harrah

H. Jack Geiger, M.D. is a professor and Chair of the Department of Community Medicine at the City University of New York School of Medicine. He is also a member of the Physicians for Social Responsibility. Tony Harrah wrote his comments as a special feature for the Guardian.

Points to Consider:

1. What is the conflict of interest at the Department of Energy?

2. How is the public health threatened?

3. Why was radioactive iodine intentionally released in the atmosphere?

4. How did operations at Hanford affect the environment and the citizens of the area?

Excerpted from testimony by H. Jack Geiger before the Senate Committee on Energy and Natural Resources, Oct. 5, 1989 and Tony Harrah, "Feds Confess: Hanford as Bad as Chernobyl," **Guardian,** August 15, 1990.

We do not know the scope of the environmental and safety problems and the extent of the risks they pose to thousands of Americans.

H. JACK GEIGER[1]

The problems at our nation's nuclear weapons production facilities have convinced me of the need for major statutory changes in the responsibilities assigned to the Department of Energy and the laws that govern operations within the nuclear weapons production complex.

Conflict of Interest

There is an inherent conflict of interest at DOE: one bureaucracy has been given both the assignment of producing nuclear weapons and protecting the public health against hazards stemming from that production. It has not worked. This situation has led to a "culture of secrecy" within the department and insufficient regard for the protection of human life and the environment.

The Secretary of Energy has said publicly, in an admission I consider bold and long overdue, that for decades public health has consistently taken a back seat to weapons production. The Radiation Research Reorganization Act is an important step toward putting public health first, where it belongs. A definition of national security that is restricted to the stockpiling of nuclear weapons and that fails to include the health of the American people and the protection of the American environment is tragically flawed.

Threat to the Public Health

From a medical perspective, the problems of accidents, mismanagement, sub-standard health and safety practices, irresponsible waste disposal, and radioactive and toxic releases into the environment at Savannah River, Rocky Flats, Hanford, Fernald, and other facilities constitute a threat to the public health. I am not asserting that thousands of lives are definitely in immediate danger — it is a threat in the sense that we do not know the scope of the environmental and safety problems and the extent of the risks they pose to thousands of Americans.

---

[1] *H. Jack Geiger's comments begin here.*

HANFORD'S LEAKING TANKS

AT THE NUCLEAR WEAPONS FACILITY IN HANFORD, WASHINGTON

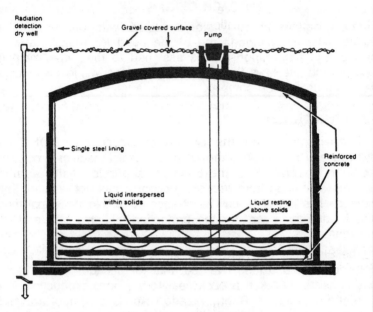

TOXIC LIQUIDS
STORED IN SINGLE-SHELLED TANKS

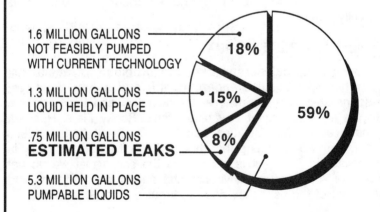

1.6 MILLION GALLONS
NOT FEASIBLY PUMPED
WITH CURRENT TECHNOLOGY — 18%

1.3 MILLION GALLONS
LIQUID HELD IN PLACE — 15%

.75 MILLION GALLONS
ESTIMATED LEAKS — 8%

5.3 MILLION GALLONS
PUMPABLE LIQUIDS — 59%

Graphics by Ron Swanson. Source: Government Accounting Office.

We are just now beginning to see how extensive these risks actually are. Published reports and disclosures of environmental and occupational health data over the last few years suggest that thousands of workers at Department of Energy nuclear weapons production facilities and people in neighboring communities may have been exposed to unacceptable, dangerous levels of radiation and toxic chemicals.

For those individuals who may have been overexposed, "secondary prevention" or early detection of disease may still be possible if the DOE will drop its cloak of secrecy and allow qualified outside experts to help. The health of many people depends on identifying the individuals or population that have been overexposed.

More important are the revelations about continuing risk of exposure in and around the DOE facilities. Radiation and toxins continue to contaminate the environment and the workplace, and people still live and work in areas that may have been contaminated in the past. Until there is clear evidence that these risks have been effectively managed, DOE and public health officials must act as if there is a continuing direct threat to people and the environment. From a medical and public health perspective, this is the only responsible course of action.

TONY HARRAH[2]

The Department of Energy (DOE) is finally coming clean about the mess at the Hanford Nuclear Reservation in southeastern Washington State. Over the past 45 years, Hanford's neighbors have been exposed to doses of radiation comparable to those given off by the Chernobyl disaster.

The department has recently begun admitting what anti-nuclear activists have known for years: that Hanford has released large amounts of radiation since it began operating in 1944 — on at least one occasion the release was intentional; that waste has been stored so carelessly it is in danger of exploding; and that the people who run Hanford have tried to hide the disaster from the public.

---

[2] *The comments of Tony Harrah begin here.*

99

NIGHTMARE AT HANFORD

The most difficult disposal challenge of all will be the 177 underground tanks, ranging in capacity from 55,476 gallons to 1.13 million gallons, containing plutonium, other radioactive elements, liquid solvents and unidentified chemicals, all mixed together in what one analyst called "a terrible cocktail".

Of these 177 tanks, 149 are singleshell tanks, consisting of a steel shell surrounded by concrete. At least 66 are known to leak.

Thomas W. Lippman, "Nuclear Nightmare Is Reality at Washington Wasteland," **Washington Post,** December 8, 1991

Inadvertent Danger

From 1944 to 1986, Hanford produced plutonium for nuclear weapons and remained a center for nuclear research. In 1986, mismanagement and mounting environmental destruction forced the shutdown of much of the complex— including the Chernobyl-style reactor that produced plutonium for nuclear warheads.

After decades of stonewalling by the Department of Energy and its predecessor, the department verified that from 1944 through the 1950s Hanford's reactors "inadvertently" released over 540,000 curies of radioactive iodine into the atmosphere. To put this number in perspective, only an estimated 15 to 24 curies were released by the partial core meltdown at Three Mile Island in 1979.

Some experts estimate that farm families living in the path of the prevailing southwest winds that sweep past the reactors could have received, over many years, radiation doses 10 times higher than people living around Chernobyl.

They Knew All Along

Solid evidence of the emissions first came to light in February 1986, when the Hanford Education Action League, a grass-roots citizens group based in Spokane, obtained over 19,000 pages of classified Energy Department documents through the Freedom of Information Act.

"My back-of-the-envelope estimate is that around 340,000 curies of iodine were released in 1945, and another 140,000 in 1946," says Bob Cook, a Richland resident and ex-employee of the Nuclear Regulatory Commission. Cook has spent years studying the documents.

"They learned how to filter it out of the stacks then, most of it; another 1,000 or so curies were released yearly after that," Cook added. "They knew even at the beginning that the stuff could be harmful, but not much else. And during the war safety wasn't the highest priority—getting plutonium made was. The smaller releases after those years came in large part from poor maintenance and carelessness, but someone knew, or should have known, what was happening. They just never told the public."

One incident doesn't fit this pattern of negligence, however. Over 7,000 curies of iodine 131 were intentionally released over the area in a December 1949 military exercise. Dubbed the "Green Run", the exercise was a test of the military's capability to monitor radioactive emissions from Hanford as far south as Klamath Falls, Oregon, near the California border.

Ed Frost, chair of World Citizens for Peace, a local anti-nuclear group, says, "The fact that the U.S. government at least once deliberately put them at risk has disappointed even many old-time Hanford workers, the ones who've always been the most defensive, the most pro-nuclear."

No Threat, Says DOE

While at last owning up to the fact of the emissions, the department still downplays their significance. It maintains, for instance, that because the iodine was released over such a long period of years, it posed no real health risk. Jim Thomas, research director for the Hanford Education Action League, says his organization disputes this claim, and is leading the effort to bring more information to light.

"A Hanford environmental dose reconstruction project was funded by Congress in 1987, under pressure from groups like us and from the state of Washington," Thomas says. He explains that the project's mission is to estimate by 1994 what areas received what doses and when, and then to determine which humans, animals and crops were present.

Another part of the dose reconstruction project's role will be to determine what other radioactive elements were released, according to Thomas, and if their effects need to be studied as well. One problem with the project's recent initial report is that it did not include the area's Native Americans. This omission especially angered the Yakima Indian Nation, which had pushed hard for the study since 1986.

Seven other Native American groups in the region could potentially be affected, and all complained of lack of resources and adequate computer models to conduct research. Ways of life that differ from those of whites and the general absence of records compound the problem of reconstructing their past. Officials with the dose reconstruction project have promised to include all eight groups in their final report.

Another hole in the study will probably remain, however. The effects of Hanford radiation on the tens of thousands of Mexican and Mexican-American migrants who worked the crops of the Yakima Valley and the Columbia Basin Irrigation Project near Hanford will likely never be accounted for.

Are there any remedies down the road for at least some of those affected? According to Hanford Education Action League's Jim Thomas, "There are class-action lawsuits in the works, but they encounter a number of problems, including the government's presumable claim of sovereign immunity," a legal doctrine under which the federal government is immune from liability for its actions.

"One likely outcome," Thomas continues, "is the dismissal of lawsuits on these grounds, and then a move for congressional compensation for victims, possibly two to four years off."

The overdue revelations from the Energy Department on iodine emissions are just the latest of Hanford's problems. In 1982, mismanagement of two atomic power plants under construction at Hanford led to the largest public bond default in U.S. history. While Bob Cook was working for the Nuclear Regulatory Commission (NRC) in 1985, he blew the whistle on evidence of massive groundwater contamination in the area, which department officials were attempting to hide.

A government advisory panel warned that the 177 radioactive waste storage tanks at Hanford were in danger of exploding. The panel reported that heat inside one or more of the tanks, or an outside spark or shock, could trigger a chemical blast spreading radiation over a large area. Millions of gallons of radioactive waste are stored at the facility.

Cleanup costs of these contamination problems at Hanford are estimated, conservatively, at $50 billion.

16 WAR AND THE ENVIRONMENT: IDEAS IN CONFLICT

NUCLEAR WEAPONS PLANTS ARE SAFE

James D. Watkins and John C. Tuck

James Watkins is the U.S. Secretary of Energy and a retired U.S. Navy Admiral. John C. Tuck is Under Secretary of the U.S. Department of Energy (DOE). Additional information was taken from the DOE's Office of Civilian Radioactive Waste Management.

Points to Consider:

1. How did Secretary Watkins propose to change DOE operations?

2. How does the Department of Energy propose to "reconfigure" our nuclear weapons complex?

3. Why is Complex-21 important?

4. What environmental safeguards will be implemented for Complex-21?

Excerpted from testimony by James D. Watkins before the Senate Committee on Energy and Natural Resources, Oct. 5, 1989, and by Robert C. Tuck before the Senate Committee on Governmental Affairs, February 25, 1991.

We can provide our vital nuclear deterrent in an environmentally safe and sound manner as we move into the next century.

JAMES D. WATKINS[1]

In an effort to instill personal accountability and responsibility quickly, effectively, and completely, and to debunk the myth that our nuclear weapons plants cannot be run in an environmentally safe manner, I have been publicly and privately tough on both the Department and its contractors. I have established a ten-point plan to turn the Department's management culture around. I have sent out "Tiger Teams" to assess the management problems in the field and I have published a five-year plan that attempts to address how we should begin fixing the environmental problems at our facilities. I have rejected the schedules set previously for DOE programs and repeatedly pushed DOE staff and contractor employees to new levels of excellence in professional performance. We are aggressively making changes.

It is my strong and sincere belief that, with continued progress in changing the management and operating culture, the system can be made to work. I have worked at keeping DOE facilities up and running, as long as they do not pose a threat to safety, health and the environment. Some of my critics think the solution to DOE's problems is to shut everything down, turn off the lights and go home. It is my adamant view, however, that we have not come to the fork in the road where our choice is either (1) unilateral nuclear disarmament of (2) full production under dangerous conditions that threaten our environment, safety and health. While the present system may be in a state of disrepair, it can be fixed. With the continued full cooperation and help of this Committee, and with a forward-looking, systematic, orderly approach, the Department of Energy can run its defense plants in an environmentally sound and safe manner and produce the material needed for the nation's defense.

Environmental Restoration

Our five-year plan is good news for you today. As I promised you earlier this year, the plan has been formulated and was published for comment. This plan provides the framework within which the Department will conduct its waste management and

[1]*The comments of James D. Watkins begin here.*

104

Nuclear, Chemical, and Biological Warfare Facilities

Nuclear Warhead Facilities:	Location:
Feed Materials Production Center	Fernald, OHIO
	Ashtabula, OHIO
Hanford Reservation	Hanford, WASHINGTON
Idaho National Engineering Laboratory	near Idaho Falls, IDAHO
Kansas City Plant	Kansas City, MISSOURI
Livermore National Laboratory	Livermore, CALIFORNIA
Los Alamos National Laboratory	Los Alamos, NEW MEXICO
Mound Plant	near Dayton, OHIO
Nevada Test Site	near Las Vegas, NEVADA
Paducah Gaseous Diffusion Plant	Paducah, KENTUCKY
Pantex Plant	near Amarillo, TEXAS
Pinellas Plant	Clearwater, FLORIDA
Portsmouth Uranium Enrichment Complex	Piketon, OHIO
Rocky Flats Plant	near Denver, COLORADO
Sandia National Laboratories	Albuquerque, NEW MEXICO
Savannah River Plant	near Aiken, SOUTH CAROLINA
Waste Isolation Pilot Plant	near Carlsbad, NEW MEXICO
Y-12 Plant	Oak Ridge, TENNESSEE

Chemical Weapons Facilities:	Location:
Aberdeen Proving Ground	Edgewood, MARYLAND
Anniston Army Depot	Anniston, ALABAMA
Johnston Atoll	Pacific Ocean
Lexington-Blue Grass Depot Activity	Richmond, KENTUCKY
Muscle Shoals Phosphate Dev. Works	Muscle Shoals, ALABAMA
Newport Army Ammunition Plant	Newport, INDIANA
Pine Bluff Arsenal	Pine Bluff, ARKANSAS
Pueblo Army Depot Activity	Pueblo, COLORADO
Rocky Mountain Arsenal	near Denver, COLORADO
Tooele Army Depot	Tooele, UTAH
Umatilla Army Depot Activity	Hermiston, OREGON
West Germany	Undisclosed

Major Biological Warfare Facilities:	Location:
Dugway Proving Ground	near Salt Lake City, UTAH
Salk Institute (Government Services Division)	Swiftwater, PENNSYLVANIA
U.S. Army Medical Research Institute for Infectious Diseases (Fort Detrick)	Frederick, MARYLAND

Sources: DOD, DOE

Prepared by Center for Defense Information.

RESEARCH PROGRAM

In response to what they perceived as crucial environmental problems, several members of the Senate Armed Services Committee introduced legislation to create a Strategic Environmental Research Program (SERP). As enacted, the legislation directs resources of DOD and DOE (the defense programs) to environmental concerns, based on the premise that environmental deterioration threatens domestic and international security and causes certain economic and political problems.

DOD and DOE already undertake environmental protection activities based primarily on their ownership and operation of many major hazardous and nuclear waste sites. Other current programs include energy conservation at military bases, pollution prevention, and waste minimization. SERP will build on existing DOD and DOE efforts, setting up an interagency council to coordinate existing programs and make recommendations.

Excerpted from testimony by Robert C. Tuck, before the Senate Committee on Governmental Affairs, February 25, 1991

cleanup activities. It is designed to address all environmental restoration and waste management requirements, whether statutory, regulatory, or by agreement. For the first time, the Department is planning a cleanup in a unified fashion and in the open, with broad participation of interested parties outside the Federal Government. Also for the first time, cleanup planning is taking place without deference to other program interests. The activities under the five-year plan will be treated as a single, focused, and unified program with a new budget and reporting structure so that Congress will be able to better oversee the progress we make in this area. Unlike so much of what has gone before, the five-year plan will be a living document, one in which our plans and priorities will be revisited each year with full public participation.

We are now in a different era, one in which a concern for environment is coupled directly with meeting other national security objectives. It is obvious to me, and I hope to you, that the Department cannot solve all of these problems alone. As I said before, I need your assistance and counsel as we all seek to resolve these problems and to maintain this nation's future nuclear deterrent capability in a manner that does not do violence to the environment or compromise public health and safety.

JOHN C. TUCK[2]

The Department of Energy has the responsibility for nuclear weapons research, design, development, testing, production, retirement, and disposal. The Department is also responsible for weapons safety, as well as the security of the weapons complex and nuclear weapons not under the control of DOD.

Nuclear deterrence has been a cornerstone of our national security policy since the end of World War II and has been instrumental in preserving world peace for the past 45 years. For the foreseeable future, it will remain an essential element of our national defense strategy.

As far back as 1987, Congress recognized that a comprehensive rather than piecemeal approach was needed to address these problems. In the National Defense Authorization Act for fiscal years 1988 and 1989 (Public Law 100-180), Congress directed that the President conduct a study and prepare a plan for modernizing the complex, taking into account the overall size, and productive capacity necessary to support national security objectives. The plan was required to ensure safe and environmentally acceptable operation and to set forth a schedule for and the estimated costs of implementation of the plan. The "Nuclear Weapons Complex Modernization Report" (Modernization Report), was submitted to Congress in January 1989 in response to this requirement.

New Considerations

When Admiral Watkins reviewed the report shortly after he became the Secretary of Energy, he concluded that the report did not incorporate significant changes in departmental policies or take into account significant international events that were unfolding.

In August 1990, in light of significant changes in the world situation, particularly with respect to the Soviet Union and Eastern Europe and the consequent likelihood of reductions in the nuclear weapons stockpile, Admiral Watkins redirected the study to give full consideration to a future complex that would be smaller, less diverse, and less expensive to operate than it is today. This redirected study was given the more appropriate title of "reconfiguration" rather than "modernization".

[2]*The comments of John C. Tuck begin here.*

The Committee's formal new look at the complex, called "The Complex Reconfiguration Study", was recently submitted to the Congress.

Goals for the Nuclear Weapons Complex of the Future

Complex-21, as this reconfigured complex has come to be known, is envisioned as the complex that will sustain the nation's nuclear deterrent capability through the middle of the 21st century. In addition to being smaller, less diverse, and less expensive to operate, it must be safe, reliable, efficient, and have the flexibility to support the nuclear deterrent under changing requirements set by the President in the nuclear weapons stockpile memorandum. The number and size of weapons production sites will be minimized, and, where practicable, non-nuclear production will be transferred to the private sector.

It is essential that the restructured weapons complex comply with applicable laws and regulations, and incorporate the highest standards regarding nuclear and industrial safety, safeguarding the health of the public and employees, protecting the environment, and ensuring the safety and security of nuclear materials and weapons components. The use of hazardous materials and the number and sizes of waste streams will be minimized. Proper disposal of hazardous, radioactive, and mixed wastes will be accommodated. In this regard, the Office of Environment, Safety and Health and the Office of Environmental Restoration and Waste Management have actively participated in the development of the complex reconfiguration study.

Department of Energy (DOE) Activities

Secretary Watkins has repeatedly emphasized to Congress, the President, and the American people his personal commitment to insure that the DOE will operate its facilities in a safe and environmentally responsible manner. The Secretary announced in January 1990 that the DOE would prepare a study on modernizing its nuclear weapons complex. We have begun that process.

Transition Activities

During the transition to a reconfigured complex, it is necessary that DOE continue to devote significant resources over the next several years to bring the existing complex into compliance with applicable federal, state and local laws, regulations, and orders and to assure that it can carry out its defense mission until Complex-21 is ready.

It is obvious that the Department must operate its existing facilities safely and reliably to meet its defense mission. The investments must be borne to upgrade and sustain these facilities. Transition activities include the following:

1. Improving safety and health conditions in full compliance with all laws and regulations;

2. Addressing and resolving environmental protection and waste management issues;

3. Improving safeguards and security;

4. Upgrading and renewing existing facilities to restore needed operational capabilities;

5. Closing and cleaning up those facilities no longer needed; and

6. Raising the maintenance tempo to reduce the backlog of deferred maintenance and to prevent further deterioration of facilities.

Activities will require extensive use of architectural, engineering, environmental and technical support to prepare and evaluate design alternatives and the associated environmental impact analyses. The reconfiguration project office will oversee these activities in close coordination with the National Laboratories and other Department offices and oversight groups.

Conclusions

The vision of Complex-21 will not be fulfilled overnight. Under Admiral Watkins' leadership, we have established a blueprint which can lead to a complex that would be smaller, less diverse and less expensive to operate than it is today. To complete it will require the continuing support of current and future administrations and Congress. Together, we can preserve the Department's capabilities to provide for our vital nuclear deterrent in an environmentally safe and sound manner as we move into the next century.

17 WAR AND THE ENVIRONMENT: IDEAS IN CONFLICT

STORING NUCLEAR WASTE: THE POINT

U.S. Department of Energy and Cecil D. Andrus

The U. S. Department of Energy (DOE) is responsible for the disposal of the nation's radioactive wastes. Cecil D. Andrus is Governor of the State of Idaho.

In 1979, Congress authorized the U.S. Department of Energy to proceed with construction of the **Waste Isolation Pilot Plant** *(WIPP) in southeastern New Mexico for the "express purpose of providing a research and development facility to demonstrate the safe underground disposal of radioactive wastes" resulting from U.S. defense activities.*

This was followed by passage of the **Nuclear Waste Policy Act** *(NWPA) in 1982, which further directed DOE to site, design and construct a permanent geologic repository for the nation's radioactive wastes. In 1987, Congress designated* **Yucca Mountain, Nevada** *for this purpose. Both sites continue to be the source of much national debate on the disposal of radioactive waste.*

Points to Consider:

1. What is the NWPA and why is it important?

2. How has DOE worked to ensure that the Waste Isolation Pilot Project (WIPP) and Yucca Mountain underground storage sites will be environmentally safe?

3. Why does Governor Andrus support the WIPP site?

Excerpted from a public statement published by the U.S. Department of Energy's Office of Civilian Radioactive Waste Management and from testimony submitted by Cecil D. Andrus before the House Committee on Interior and Insular Affairs, April 16, 1991.

It is time to make a meaningful commitment to begin cleaning up this nation's nuclear waste problems.

The safe and permanent disposal of high-level nuclear waste is a major objective of the United States, as well as many other nations. [1]The Federal Government is responsible for the disposal of this waste. The waste will be disposed of in a deep underground repository to isolate the waste from the public and the environment and to prevent human intrusion into the repository.

Since the nuclear age began more than 40 years ago, high-level nuclear waste has been accumulating. This waste includes spent, or used, nuclear fuel from commercial powerplants and high-level radioactive waste from defense activities. U.S. commercial spent fuel in temporary storage maintained by utilities totaled approximately 18,000 metric tons at the end of 1988. Approximately 8,000 to 9,000 metric tons of defense high-level waste are currently stored temporarily at three U.S. Department of Energy (DOE) sites. While waste has been safely stored for decades, it will be radioactive for thousand of years and, therefore, requires permanent disposal.

NWPA

Passage of the *Nuclear Waste Policy Act* of 1982 (NWPA) was a major milestone in the Nation's management of high-level nuclear waste. Signed into law by the President on January 7, 1983, and amended by the Nuclear Waste Policy Amendments Act of 1987 on December 22, 1987, this legislation established a national policy for safely storing, transporting and disposing of spent nuclear fuel and high-level radioactive waste. The NWPA established the Office of Civilian Radioactive Waste Management (CRWM) within the U.S. Department of Energy (DOE) to implement the policy and to develop, manage and operate a safe waste management system to protect public health and the environment. The major responsibilities assigned to DOE include the following:

- To site, construct and operate a deep, mined geologic repository
- To site, construct and operate one monitored retrievable storage (MRS) facility, under certain conditions
- To develop a system for transporting the waste to a repository and an MRS facility.

[1] *The comments of the U.S. Department of Energy begin here.*

Surface facilities at the repository will be located about one mile east of the underground facilities.

Source: U.S. Department of Energy.

Nuclear waste is the byproduct that results from using radioactive material. Because it is radioactive, disposing of it requires special handling to avoid the health and environmental hazards associated with radiation. There are four categories of nuclear waste: high-level waste, low-level waste, transuranic waste and mill tailings. Classification of waste depends on its origin, level of radioactivity and potential hazard.

High-Level Waste

High-level waste, such as spent fuel from nuclear power plants and the waste from defense activities, is the most radioactive category of nuclear waste. It usually decays or loses radioactivity rapidly. However, some high-level waste may also contain elements that decay very slowly and may remain radioactive for thousands of years. Most high-level waste must be handled by remote control from behind heavy protective shielding.

Under provisions of the Nuclear Waste Policy Act of 1982 and the Nuclear Waste Policy Amendments Act of 1987, the U.S. Department of Energy (DOE) is responsible for establishing a system for the disposal of high-level radioactive waste in a deep underground facility known as a repository. The Amendments Act directed DOE to perform an intense study, known as site characterization, on Yucca Mountain, Nevada, as the candidate site for the geologic repository.

Low-Level Waste

Low-level waste typically contains a small amount of radioactivity dispersed in a large amount of material and poses little potential hazard. Low-level waste results from many commercial, medical and industrial processes. It includes items such as rags, papers, filters, resins and discarded protective clothing from "housekeeping" functions of commercial and university nuclear facilities. Low-level waste does not require extensive shielding, but some protective shielding may be needed for handling certain low-level waste. Commercial low-level waste is currently disposed of by shallow land burial in containers at three federally licensed sites: *Barnwell, South Carolina; Beatty, Nevada; and Hanford, Washington.*

Under provisions of the amended Low-Level Radioactive Waste Policy Act of 1980, by January 1, 1993, each state will be responsible for ensuring safe disposal of low-level waste produced commercially within its borders or may enter into regional contracts with other states to provide for disposal of low-level waste from all compact members at a regional facility. These compacts must be approved by Congress.

Transuranic Waste

Transuranic waste consists of manmade elements formed as a by-product of reactor operation. It emits less penetrating radiation than high-level waste and decays, or loses radioactivity, slowly. Its total radioactivity may be no greater than certain low-level waste. Most transuranic waste results from reprocessing nuclear fuel as a part of the nation's defense activities.

Some transuranic waste is now being stored in surface facilities, but current plans call for most of this waste to be placed in a deep, geologic repository because, like high-level waste, it remains hazardous for long periods of time while decaying.

Mill Tailings

Mill tailings are naturally radioactive rock and soil that are by-products of mining and milling uranium. They contain small amounts of radium that decays to radon, a radioactive gas. Plans are being developed by DOE for controlled disposal of mill tailings at isolated locations where they will be covered with enough soil to protect the public and the environment from the radon associated with the mill tailings.

Monitored Retrievable Storage (MRS)

Under the provisions of the NWPA, as amended, DOE is authorized to site, construct and operate an MRS facility, subject to certain conditions. An MRS facility would receive, package and temporarily store spent fuel from nuclear power plants before transporting the fuel to the permanent geologic repository by special trains. The DOE considers an MRS facility to be a key component of the nuclear waste management system.

WASTE PILOT PROGRAM (WIPP)[2]

As Governor of the State of Idaho, I have a very strong motivation to see that the WIPP facility is opened in a timely fashion. Idaho is home to thousands of barrels of transuranic (TRU) waste which have been dumped into the ground for years at the Idaho National Engineering Laboratory (INEL). This "storage" site in Idaho has been listed on the Environmental Protection Agency's National Priorities List as a "Superfund" site. As long as the TRU waste remains buried in the ground at the INEL, it is a threat to Idaho's people and to Idaho's most important groundwater resource, the Snake River Plain Aquifer

It is time to make a meaningful commitment to begin cleaning up this nation's nuclear waste problems—the opening of the WIPP facility is a major positive step in that direction. As I have said many times, we need to put as much, if not more, emphasis on the back end of nuclear technology as we have on the front end. The opening of WIPP sends a strong signal that both the administration and Congress are ready to step up to this task.

---

[2] *The comments of Gov. Cecil Andrus begin here.*

The WIPP facility has been fully scrutinized and studied to ensure that it meets environmental concerns and the Environmental Protection Agency's no-migration determination. The independent Blue Ribbon Panel that was appointed to assess issues regarding the WIPP facility has recommended that the test phase of the facility begin. The tax payers of this nation have invested nearly one billion dollars in studying and developing the facility. With this as a backdrop, WIPP stands ready to begin receiving TRU waste that is presenting significant health, safety and environmental risks in the western states.

I understand that there are some technical issues that remain to be reviewed. I trust that the Committee will not become so engrossed by the details of the remaining technical issues that it loses sight of the objective of this legislation—to open the WIPP facility for a five-year test phase.

Under any scenario, only a very small percentage of TRU waste that is now dumped in various locations throughout the west will be placed in the WIPP facility during this test phase. But it is time to begin. It is time for Congress as a whole to become significant players in the resolution of our nation's nuclear waste problems by passing legislation to allow the WIPP facility to open.

18 WAR AND THE ENVIRONMENT: IDEAS IN CONFLICT

STORING NUCLEAR WASTE: THE COUNTERPOINT

Michele Merola

Michele Merola is the Executive Director of Concerned Citizens for Nuclear Safety, a non-profit public education organization based in Santa Fe, New Mexico.

Points to Consider:

1. Why is the Waste Isolation Pilot Project (WIPP) scientifically unsound?

2. What is wrong with underground storage of nuclear waste?

3. Why is "clean-up" a myth?

4. What other options are proposed by the author?

Excerpted from testimony by Michele Merola before the House Subcommittee on Energy and the Environment of the Committee on Interior and Insular Affairs, April 16, 1991.

It is arrogant of us to believe that we have perfected a waste burial repository which will safely shield the environment from these radioactive materials for a quarter of a million years.

The Waste Isolation Pilot Project (WIPP) is intended to set the precedent for safe nuclear waste disposal in this country. As such, WIPP should be the very model of prudent scientific planning, and all environmental, health and safety laws should be strictly adhered to in the course of this project's development. But unfortunately, so far WIPP has failed to meet any of these criteria. The history of WIPP is marked by unexpected geologic flaws, aborted certification attempts, legal variances and exemptions, and the complete failure of the DOE to justify its plan to "test" radioactive waste underground for five to seven years before complying with Environmental Protection Agency standards for the disposal of transuranic radioactive waste. (Transuranic waste consists of manmade elements formed as a by-product of reactor operation.)

The DOE itself has characterized its schedules for WIPP in 1989 as "fallacious and scientifically unsound". It is regrettable that this description remains accurate today.

Is WIPP the Answer?

It is important to recognize that even if WIPP were to open for a test phase, it is not going to relieve the nuclear waste problem. Fewer than 5000 barrels of waste would be sent to WIPP over the next seven years under the current plan. Even at full capacity, WIPP would hold only 1.3% by volume of the radioactive waste currently in existence. And approximately 70% of WIPP's volume is reserved for nuclear weapons waste yet to be generated. It appears that the only real value of rushing WIPP's opening is its public relations value: by starting the shipments of waste to WIPP, the DOE would instill in the public mind the illusion that there is a solution for this nation's radioactive waste dilemma. But WIPP's obvious shortcomings demonstrate that despite the possible PR value of the project, the very real problem of how to "dispose" of nuclear waste will still be unaddressed.

Antiquated Disposal Technology

The technological advances over the 35 years since the idea behind WIPP was first conceived now show us how primitive

118

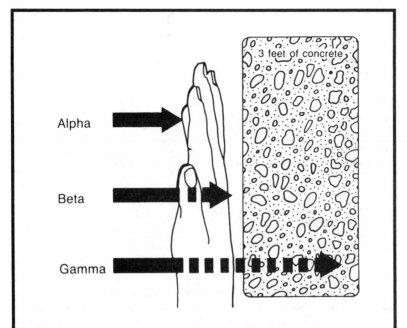

Alpha

Beta

Gamma

3 feet of concrete

Three types of radiation are associated with using nuclear energy and creating nuclear waste, and all three are also found in nature. While all three types of radiation are potentially harmful, they differ in their penetrating power and in the way they affect human tissue.

Source: U.S. Department of Energy

and ultimately irresponsible the idea of simply burying nuclear waste really is. The waste intended for WIPP is contaminated with plutonium, a man-made element which will remain radioactive and lethal for more than 240,000 years. It is arrogant of us to believe that we have perfected a waste burial repository which will safely shield the environment from these radioactive materials for a quarter of a million years.

Los Alamos National Laboratory is currently researching a process utilizing a linear accelerator which has shown promise in greatly reducing the half-lives of radioactive materials through transmutation of the elements. The irrevocable burial of radioactive waste at WIPP would preclude the use of any possible future treatment options, such as transmutation, which may be developed in the next few decades. In fact, a 1981 study by Los Alamos National Laboratory rated deep geologic burial dead last out of 14 possible waste management options under consideration.

A GRADE OF "F"

Each of us who lives or works or goes to school along a WIPP route, or who comes neas a trupact truck on the highway, will be exposed again and again and again, until the radiation build-up produces cancer in our bodies.

The Dept. of Transportation guidelines stress that "the state adequately consider public risk to all those who may be affected by radioactive material transportation." We the public have decided that transporting the waste cannot be made sufficiently sage and the risk is too great for any route that may be considered.

DOE wants to test the WIPP site and the transportation process. I am a mathematics instructor. Sometimes I teach a course in logic. The so-called logic of wanting to test something that is already known to be unsafe would draw a grade of "F" in any college logic course!

Another word for test is experiment. Each situation in which the U.S. government has experimented with radiation in the past, people have contracted and died from cancer. The prople experimented upon have been minorities, rural people, or servicemen who ended up being punished for serving their country.

Almost every hazardous waste dump around the nation is in an area where higher concentrations of minorities reside. We do not want to be the next group of minorities and rural people to be the government's guinea pigs.

Excerpted from testimony submitted by Penelope McMullen, Peace and Health Coordinator, Sisters of Loretto in testimony before the Senate Committee on Energy and Natural Resources, September 21, 1991

The Myth of "Cleanup"

"But nuclear waste has to go somewhere." This is the claim that has been driving both WIPP and the Yucca Mountain projects for so long. However, the DOE itself concedes that the waste intended for WIPP can be safely stored as it is now with minimal adverse impacts. Although the DOE often touts WIPP as a vital component of "cleanup", WIPP is intended to hold only mid-level transuranic waste currently held in retrievable storage (usually in 55 gallon drums on concrete pads). This retrievably-stored TRU waste is probably the only form of radioactive waste in the nuclear weapons complex that is not currently a threat to the environment. There is no plan to dig up contaminated areas or otherwise "clean up" contaminated DOE sites and send the waste to WIPP.

The environmental degradation witnessed at every single DOE nuclear weapons facility is the result of years of mismanagement, neglect and just plain ignorance. Prior to 1970, nuclear waste was dumped into pits, ponds and trenches buried in cardboard boxes, and even intentionally injected into deep wells. Indeed, some of these practices continue today. This is the waste that is currently posing the greatest threat to both the environment and human health. But this waste is the subject of a separate Environmental Restoration and Waste Management plan; a series of hearings were just held across the country on DOE's proposal to clean up the thousands of sites contaminated by its nuclear weapons operations. WIPP does not alleviate the problems from this waste which has already been dumped into the environment. Nor will it address such pressing problems as the potentially explosive high-level waste in the Hanford tanks.

The immediate effect of WIPP would be to relieve storage space problems at Idaho and Rocky Flats. The stored transuranic waste at these sites is the waste scheduled to be sent to WIPP. WIPP's primary purpose is to provide for the continuation of nuclear weapons production through the year 2013. As previously mentioned, 70% of WIPP's capacity is reserved for waste that doesn't even exist yet. And with changes in world superpower relationships, the need to continue producing nuclear weapons at current levels is debatable. The need to open WIPP immediately, before it is proven to be safe, is questionable at best.

Alternate Course of Action

The geological failings of WIPP and the potential advances of technology have rendered burial of nuclear waste an outdated, unsafe and irresponsible course of action. Given the rapid advances of science and the valuable time saved by on-site storage of waste, it would make good sense to store the waste at the generator sites simultaneous with a major research effort into viable methods of waste treatment and/or transmutation. In fact, the DOE's *Nuclear Weapons Complex Reconfiguration Study* recommends that weapons production facilities provide for on-site storage of radioactive waste. (WIPP is specifically excluded, however.) If waste is buried at WIPP now, the radioactivity we have created will be out of our control for thousands of years, and that waste will no longer be available for treatment options which may be developed within the next few decades.

Just think of the scientific advances made in the last 35 years. Now think of the possibilities in the next 35. Since the waste that is scheduled to go to WIPP is not a threat to the environment and can be stored safely on site, it just doesn't make sense to eliminate future options by burying nuclear waste at WIPP. We need to give waste management technologies a chance to catch up with the technologies which generated this waste in the first place.

National Scope

People in New Mexico opposed to WIPP are often accused of displaying the NIMBY syndrome (Not In My Back Yard). This may be true in many cases, because we do value the natural beauty and pristine environment of New Mexico, and the indigenous people here have a history and reverence for their native land which is unparalleled in the U.S. But the problem goes far deeper than that. The handling of nuclear weapons waste is a matter of responsibility. We reject the "out of sight, out of mind" mentality represented by WIPP. Our technology has created deadly radioactive elements which last far longer than most of us are capable of conceptualizing. Now we have to develop a technology to deal responsibly with the long-lived dangers of this waste, not just sweep it under the rug in New Mexico only to have it resurface at some later date.

Support for on-site storage of nuclear weapons waste is national in scope. In September 1990, the Military Production Network, a coalition of more than 40 national, regional and grass-roots organizations working on issues of nuclear weapons production and waste management, issued a statement in favor of on-site storage and in opposition to non-solutions such as WIPP and Yucca Mountain. These were groups from extensively contaminated sites, such as Hanford, Rocky Flats and Savannah River.

People who are familiar with the issues recognize that WIPP is only a shell game for nuclear waste, and does not address the root of the problem. Here in New Mexico, we have two major DOE facilities, Los Alamos and Sandia National Laboratories. At Los Alamos, just 26 miles from Santa Fe, there is far more waste currently held on-site than at many of the other facilities slated to send waste to WIPP, including Rocky Flats. We support the continued storage of this waste at Los Alamos simultaneous with research into responsible, alternative technologies to manage this waste safely over time so that we do not irrevocably contaminate the environment for our children. Burying this waste at WIPP only foists the problem off onto future generations.

What Are the Real Costs of WIPP?

The cost of operating WIPP over 25 years is projected to be more than $2.5 billion; already, nearly $1 billion has been spent on WIPP. With every indication that WIPP is not only an unsafe, but also an anachronistic approach to nuclear waste disposal, wouldn't it be best to cut our losses now and save the taxpayers from throwing good money after bad? We can still save ourselves at least $1.5 billion (probably more given the DOE's history of 500% cost overruns) and eliminate both the risks and costs of a massive transportation program if the waste is stored on-site. The savings could go toward alternative waste treatment and/or transmutation research programs, and the waste could be stored safely on-site while these methods are pursued.

In addressing the very real problem of nuclear waste generated by the DOE weapons complex, long-term environmental protection must not be sacrificed for the sake of short-term public relations benefits. As it stands, WIPP and the Yucca Mountain Project are monuments to the U.S. Department of Energy's PR efforts rather than technical excellence. Geologic flaws and attempts to sidestep federal environmental laws have been glossed over with soothing assurances from an agency with no credibility and a hidden agenda.

The U.S. Department of Energy is allegedly attempting to mend its negligent ways, but WIPP has continued unabated in spite of mounting scientific evidence that the site cannot safely isolate radioactivity from the environment for the vast time spans required. The people of New Mexico do not believe that the burial of nuclear waste — burial anywhere — is a responsible method for addressing the problem. We advocate storage of nuclear waste at the generator sites, simultaneous with a major research effort into alternative methods of waste treatment and/or transmutation.

If WIPP is to continue, however, it must at least be done right. Legal variances and exemptions do not instill public confidence, do not provide a solid scientific foundation for the project, nor do they set a sound precedent. Fortunately, Congress has regained control of the land transfer at WIPP. Through land withdrawal legislation, Congress has the tool for ensuring that all environmental, health and safety laws are met at WIPP and that all outstanding issues are resolved before any waste is brought to the site for testing or otherwise.

INTERPRETING EDITORIAL CARTOONS

This activity may be used as an individualized study guide for students in libraries and resource centers or as a discussion catalyst in small group and classroom discussions.

Although cartoons are usually humorous, the main intent of most political cartoonists is not to entertain. Cartoons express serious social comment about important issues. Using graphic and visual arts, the cartoonist expresses opinions and attitudes. By employing an entertaining and often light-hearted visual format, cartoonists may have as much or more impact on national and world issues as editorial and syndicated columnists.

Points to Consider:

1. Examine the cartoon on the next page.

2. How would you describe the cartoon's message?

3. Try to summarize the message in one to three sentences.

4. Does the cartoon's message support the author's point of view in any of the opinions in Chapter Three of this publication? If the answer is yes, be specific about which reading or readings and why.

BIBLIOGRAPHY

GENERAL

Adler, J. Survival. *Newsweek,* v. 116, Dec. 31, 1990: p. 30.

Bahiri, S. Ecological safe manufacturing and Israeli conversion. *The New Economy,* Feb. 1991: p. 8.

Baker, B. Testing the waters. *Common Cause Magazine,* v. 17, Jan/Feb 1991: p. 22-27.

Bernstein, D. Downwinders study past, worry about future. *National Catholic Reporter,* v. 26, Sept. 18, 1990: p. 1.

Blair, B. G. Accidental nuclear war. *Scientific American,* v. 203, Dec. 1990: p. 53-58.

Bloom, S. Base closures and cleanups: The hidden costs of militarism. *The New Economy,* v. 2, Summer 1991: p. 4-5.

Bukowski, G. The militarization of Nevada. *Earth Island Journal,* Spring 1990: p. 20-21.

Cochran, T. B. A first look at the Soviet bomb complex. *The Bulletin of Atomic Scientists,* v. 47, May 1991: p. 25-31.

Covino, C. P. Global materialism threatens planet. *USA Today* v. 119, Jan. 1991: p. 6-7.

Chow-Bush, V. The fallen angel. *Scholastic Update,* v. 123, Apr. 19, 1991: p. 22-23.

Danielson, B. French slam "open door" on Greenpeace. *The Bulletin of Atomic Scientists,* v. 47, March 1991: p. 6-7.

Deudney, D. Environment and Security. *The Bulletin of Atomic Scientists,* v. 47, April 1991: p. 22-28.

Ecology of war and peace. *Natural History,* Nov. 1990: p,. 34-49.

Emerson, S. When Earth takes the hit. *International Wildlife,* v. 21, Jul/Aug 1991: p. 38-41.

Fuhrman, P. Ammo dump, anyone? *Forbes,* v. 146, Oct. 15, 1990: p. 40-41.

Gleick, P. H. Environment and security. *The Bulletin of Atomic Scientists,* v. 47, April 1991: p. 16-21.

Hamilton, M. Papered over [Rocky Flats, Colorado]. *Mother Jones,* v. 15, Nov/Dec 1990: p. 29-30.

Harris-Monin, F. Celestial junkyard. *World Press Review,* v. 38, May 1991: p. 54.

Harvey, H. Rethinking national security. *Utne Reader,* Jan/Feb 1990: p. 36.

Kopecky, G. The way we were. *Redbook,* v. 176, March 1991: p. 108.

Krauskopf, B. Disposal of high-level nuclear waste: is it possible? *Science,* v. 249, Sept. 14, 1990: p. 1231-2.

Krugman, P. R. The economic history lessons of war. *U.S. News and World Report,* v. 110, Jan. 28, 1991: p. 51.

Lanouette, W. J. Savannah River halo fades. *The Bulletin of Atomic Scientists,* v. 46, Dec. 1990: p. 26-29.

Laurin, F. Scandanavians underwater time bomb. *The Bulletin of Atomic Scientists,* v. 47, March 1991: p. 10-15.

Linden, E. Is the planet on the back burner? *Time,* v. 138, Dec. 24, 1990: p. 48-50.

Mardon, M. The military and the sublime. *Sierra,* v. 76, May/June 1991: p. 79-81.

Marquez, G. Four tons of dynamite for everyone! *Utne Reader,* Jan/Feb 1990: p. 123-5.

McAlevey, J. F. El Salvador's new government escalates the war against the land. *Earth Island Journal,* Winter 1990: p. 45.

Miles, S. Lessons of El Salvador. *The Nation,* v. 252, May 27, 1991: p. 698-700.

Misrach, M. W. Nerve gas unnerves Hawaiians. *Sierra,* v. 76, Mar/Apr 1991: p. 68-69.

Misrach, M. W. The Pentagon versus Hawaii. *The Progressive,* v. 55, July 1991: p. 16.

Misrach, R. Bravo 20: The Bombing of the American West. *Johns Hopkins Univ. Press.*

Root-Bernstein, R. S. Infectious terrorism. *The Atlantic,* v. 267, May 1991: p. 44.

Shulman, S. Pentagon pollution. *Technology Review,* v. 94, July 1991: p. 13-14.

Slovic, P. Lessons from Yucca Mountain. *Environment,* v. 33, April 1991: p. 6-11.

Space junk: Fears and fallacies. *Sky and Telescope,* v. 81, June 1991: p. 581.

Steele, K. D. Hanford in hot water. *The Bulletin of Atomic Scientists,* v. 47, May 1991: p. 7-8.

Stix, G. Less of a problem? *Scientific American,* v. 263, Oct. 1990: p. 124.

Sullivan, A. Fizzling out [Kuwait's oil fields]. *Wall Street Journal,* April 26, 1991: p. 1.

Turque, B. The military's toxic legacy. *Newsweek*, v. 116, Aug. 6, 1990: p. 20-23.

War and the environment. *Audubon*, v. 93, Sept/Oct 1991: p. 88-96.

PERSIAN GULF WAR

Ahlberg, B. Operation Earth Storm. *Utne Reader*, May/June 1991: p. 30-32.

Clouds Over Kuwait. *World Press Review*, v. 38, May 1991: p. 55.

Hoffman, M. Taking stock of Saddam's fiery legacy in Kuwait. *Science,* v. 253, Aug. 30, 1991: p. 971.

Holden, C. Kuwait's unjust deserts. *Science*, v. 251, Mar. 8, 1991: p. 1175.

Kaku, M. War and the environment. *Audubon,* v. 93, Sept/Oct 1991: p. 91-93.

Lacoya, R. A war against the earth. *Time,* v. 137, Feb. 4, 1991: p. 32-33.

Linden, E. Getting blacker every day. *Time,* v. 137, May 27, 1991: p. 50-51.

McGowan, A. The complete cost of war. *Environment,* v. 33, April 1991: p. 1.

Negin, E. Ecocide in the Gulf. *Scholastic Update* (teacher's edition), v. 123, April 19, 1991: p. 8-9.

Nossiter, B. D. Sand dollars. *The Progressive*, v. 55, April 1991: p. 24-27.

Pope, C. War on Earth. *Sierra,* v. 76, May/June 1991: p. 54-58.

Sheets, K. R. Iraq's environmental warfare. *U.S. News and World Report,* v. 110, Feb. 4, 1991: p. 60.

Stover, D. Shock therapy [Kuwait oil well fires]. *Popular Science,* v. 239, Oct. 1991: p. 33.

Travis, J. A legacy of war. *Science News*, v. 140, July 13, 1991: p. 24-26.

Wallis, J. The forgotten moral issues. *The Progressive*, v. 55, March 1991: p. 30-32.

War in the Gulf. *Newsweek*, v. 117, Feb. 18, 1991: p. 26-32.

Warner, F. The environmental consequences of the Gulf War. *Environment,* v. 33, June 1991: p. 6-9.

What do you think? *Scholastic Update* (teacher's edition), v. 123, May 3, 1991: p. 16-18.

GLOSSARY OF TITLES

AFB	U.S. Air Force Base
ASAT	Anti-satellite weapons
BLM	Bureau of Land Management
CBW	Chemical/biological warfare
CIA	Central Intelligence Agency
DOD	Department of Defense
DOE	Department of Energy
ECTC	Electronic combat test capability
EPA	Environmental Protection Agency
GEERT	Gulf Environmental Emergency Response Team
GAO	Government Accounting Office
ICBM	Inter-continental ballistic missile
NASA	National Aeronautical Space Administration
NEPA	**National Environmental Policy Act (of 1969)**
NPR	New Production Reactor
PEIS	Programmatic Environmental Impact Statement
R&D	Research and development
RTG	Radioactive thermal generator
SDI	Strategic Defense Initiative (Star Wars)
SERP	Strategic Environmental Research Program
Trident	(Advanced class of nuclear submarine)
UN	United Nations
USAID	U.S. Agency for International Development
USNSC	U.S. National Security Council
WIPP	Waste Isolation Pilot Plant (New Mexico)

APPENDIX

International Physicians for the Prevention of Nuclear War (IPPNW)
126 Rogers St.
Cambridge, MA 02142

Nevada Cattlemen's Association
419 Railroad St.
Elko, NV 89801

Political Ecology Group
519 Castro St., Box 111
San Francisco, CA 94114

Radioactive Waste Campaiagn
625 Broadway, 2nd Floor
New York, N Y 10012

U.S. Department of the Army
Office of the Assistant Secretary
Washington, D.C. 20310-0103

U.S. Department of Defense (USDOD)
Office of Press Secretary
Washington, D.C. 20005

U.S. Department of Energy (DOE)
Office of Press Secretary
Washington, D. C. 20005

U.S. Department of the Interior
Bureau of Land Management
Washington, D.C. 20240

World Watch Institute
1776 Massachusetts Ave., NW
Washington, D.C, 20036

**ALSO WRITE TO YOUR ELECTED OFFICIALS
IN WASHINGTON:**

Senator _____
U.S. Senate
Washington, D.C. 20510

Representative _____
U.S. House of Representatives
Washington, D.C. 20515